Snow

A History of the World's Most Fascinating Flake

Anthony R. Wood

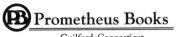 Prometheus Books

Guilford, Connecticut

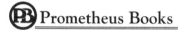
Prometheus Books

An imprint of The Rowman & Littlefield Publishing Group, Inc.
4501 Forbes Blvd., Ste. 200
Lanham, MD 20706
www.rowman.com

Distributed by NATIONAL BOOK NETWORK

British Library Cataloguing in Publication Information Available

Library of Congress Cataloging-in-Publication Data

Names: Wood, Anthony R., 1948– author.
Title: Snow : a history of the world's most fascinating flake / by Anthony R. Wood.
Description: Amherst, N.Y. : Prometheus Books, 2019. | Includes bibliographical references and index. | Summary: "The complete story of snow, this is the first book to fully examine snow as a historical, cultural, and scientific phenomenon"— Provided by publisher.
Identifiers: LCCN 2019014712 (print) | LCCN 2019981090 (ebook) | ISBN 9781633885943 (hardcover) | ISBN 9781633885950 (ebook)
Subjects: LCSH: Snow. | Snow—United States. | Snow—Forecasting.
Classification: LCC GB2603.7 .W66 2019 (print) | LCC GB2603.7 (ebook) | DDC 551.57/84—dc23
LC record available at https://lccn.loc.gov/2019014712
LC ebook record available at https://lccn.loc.gov/2019981090

Contents

Acknowledgments

During my long career at the *Philadelphia Inquirer* I have had the good fortune to work for editors who encouraged me to indulge my lifelong fascination with weather and share my perspectives on a phenomenon as fascinating as human behavior.

Along the way I have encountered some terrifically knowledgeable people who have generously shared their wisdom about the one element that has most captured my attention, and that of our readers: the magically varied crystals that we call "snow."

I would like to thank such National Weather Service meteorologists as Jim Eberwine, Tony Gigi, Chet Henricksen, and Gary Szatkowski, who have accepted the often-humbling challenges of snow forecasts and never hid when they went awry; private-sector meteorologists like AccuWeather legend Elliot Abrams; and Rutgers University snow expert Dave Robinson.

I am grateful for the New York Historical Society, the Free Library of Philadelphia, the National Park Service, the New England Ski Museum, and the NOAA Library, among other institutions, which have been resource treasuries.

Among my editors, I would especially like to thank Gene Foreman, Max King, Bill Marimow, and Vernon Loeb. *Inquirer* colleagues Mari A. Schaefer and Tommy Rowan were invaluable in assembling the images for this book. Former colleague Gilbert M. Gaul, who won two Pulitzers and deserved more, is the best journalism instructor I have ever had. And a special thank-you to Ken Libbrecht at CalTech, who helped me see snow crystals in ways I had never imagined.

All clichés aside, my gratitude to my forbearing wife, Laura, is inexpressible. How often have I subjected her to tiresome ruminations on snow these last three decades. Despite that abuse, I finished the book only through her encouragement and careful and (mostly) patient editing.

Introduction

Snow has affected the outcome of wars. It has been vital to the world economy. From the Atlantic Coastal Plain to the Sierra Nevadas to the arid West, it has been vital in the transformation of the United States from wilderness to economic superpower. It might be a last line of defense against global warming, so much so that the future of snow has become a tremendous source of anxiety in the climate community. But beyond its paramount importance as a natural resource, snow is embedded in the national consciousness in ways that transcend the practical or logical. For millions of Americans what is most compelling about snow is the complex and enchanting phenomenon itself. In a perfectly formed snowflake, a magical metamorphosis of 100,000 water droplets, a true snow-lover might see the DNA of God.

The evidence of snow's pervasive hold on so much of the public is indisputable. Watch how the weather chat boards explode with traffic when a winter storm becomes a gleam in a computer model's eye and how government and private websites become inundated with traffic when a threatened storm approaches a major population area. See how much of local TV news segments are devoted to weather when even a rumor of snow is in the vicinity. Almost nothing draws traffic to my newspaper's website like the word "snow." Watch what snow can do for swelling The Weather Channel's audience and ad revenues. The Weather Channel (TWC) gave snowstorms names, and they became at once the protagonist and antagonists in some of their most-popular running dramas.

Snow fired the curiosity of a Chinese philosopher more than 2,000 years ago and that of Johannes Kepler and other intellectual giants of the Scientific Age in Europe in the 17th century. But what is it about snow in America?

The nation has a unique and complex relationship with the six-sided crystals that Emerson described as nature's "masterpiece." One obvious reason is the fact that no other country has so many densely populated areas prone to frequently snowy winters and mega storms. Residents of the populous New York state Snow Belt routinely endure 100-inch-plus winters. The East Coast is situated near one of the planet's most volatile storm breeding grounds. Blizzards roar out of the Rockies, winter cyclones blow up along fronts in the Midwest, and blue northers drive snow into the heart of Texas.

Anthony Gigi, a wise, veteran government forecaster who moderates a popular weather forum, has observed, "Meteorology would be so much easier if everything was linear, but alas it's not and so it isn't." That axiom would apply to the nth power in the pursuit of reasoning to explain the nation's often-bifurcated relationship with snow and its mystical hold on the imagination: why some people will do almost anything to avoid it, while others pine for it and even go to great trouble and expense to make it.

The first European settlers were awed by what they encountered in the New World, what to them were literally fantastic snows and cold they had never experienced, even though they were living in the Little Ice Age, when Europe was colder than it is today. This was a climate as alien to them as the indigenous peoples they encountered. That interest in snow flourished with the Industrial Revolution and gained only more intensity with the record snows of the last three decades as the world has become warmer. I have tried to address fundamental questions that have long piqued my curiosity about the science of snow—why it snows; why it is that in the 21st century meteorologists still struggle with snow forecasts, despite having the most powerful computers taxpayers and private interests can buy; and the question of just what kind of future snow might have in a warming world. The vanishing of sun-repelling snow and ice clearly has accelerated warming in the Arctic, which in turn has led to more melting. Nothing happens in isolation in the atmosphere. Ultimately, all weather is global, and changes in the Arctic are affecting weather in the populated lower latitudes.

Snow has power that is at once menacing and magical. It can impose a cease-and-desist order on the normal business of life. It can mute the cacophony of the noisiest city with its silencing powers, while transforming the bleakest urban environment into a winter playground.

It so happened that I grew up in one of those bleak, noisy urban environments, Chester, Pennsylvania, a fading factory town on the Delaware River about 10 miles from Philadelphia. It was an industrial behemoth during the two world wars and then rapidly went the way of other places where ships and anchors, and steel and paper products, actually were once manufactured. In my youth it was in a constant state of Monday morning, a brooding ship-yard whistle dictating its rhythms and factory smoke masquerading as clouds. But a decent snowfall routed the smoke, the roaring trains inciting blizzards of white clouds along the railroad tracks. The uncanny snow animated the humble bricking of the houses and stonework of the churches. It found its way into the nooks and crevices of architectural details and everyday objects, Emily Dickinson's "alabaster wool" that fills the "wrinkles of the road." Chester, Pennsylvania, became a fantasyland. Once I discovered that snow could close school, which I viewed as a minimum-security prison with a liberal weekend furlough plan, I was hooked for life.

❋

On the subject of school, one of the highlights of my professional life occurred early in my career when I worked at United Press International (UPI) in Philadelphia. As a new recruit, one of my duties on snow days was to relay the school-closing numbers from the city government, which had agreed to be the regional collection center for the best news that a kid could possibly receive, to our broadcast clients. The numbers came via teletype, and it was my job to retype them and put them on the broadcast wire.

I embraced my awesome responsibility. How often in my youth had I begged and prayed to hear that I would be spared my scholastic sentence for a day; now here I was, emissary of the too-good-to-be-true. Tedium was never so blessed. On one morning, conservatively I had punched 300 numbers. I got 299 right. Unfortunately, the incorrect number I did type matched the code for one of the area's largest school districts. Those schools were very much open that day. The superintendent was not happy; the radio station was not happy and eventually dropped UPI. My boss was not happy. Me, I have absolutely no regrets.

❋

The realization that snow could close schools came after I had been securely captivated by snow. I recall a snowstorm on a Friday night and my brother saying it was a waste because schools would be closed anyway until Monday. I didn't feel that way at all. Snow was snow. I don't know that I was aware of it at the time, but eventually I came to appreciate that it was my invitation to the mysteries of nature, and through a passion for snow I derived invaluable lessons. Among them: You can't always get what you want and sometimes not even what you need; grown-ups were indeed fallible; and, in the words of Robert Frost, "Nothing gold can stay." I endured the heartache of anticipating a snowstorm that I hoped would close my school for life, only to get nothing—or worse, rain. Thanks to snow I discovered that even adults could be clueless. I'll never forget a TV weathercaster's saying that the precipitation the next day could be rain or snow—it all depended on whether the temperature was 32. Say what? I remembered a March storm in which a foot of snow fell and the temperature stayed above freezing the entire time.

One December morning I trudged to school through a mix of snow and rain, and a slurry of slush. The forecast said it would change to all rain and go up to 40. By midmorning, however, saucer-size flakes were accumulating rapidly. Miss Conley, our teacher, a wise and wonderful woman, looked out the window and then back to us. "Please boys and girls," she pleaded, "pray that it stops snowing." What I learned from that experience was that even an intelligent adult could spend six hours a day with nine-year-olds and know so little about them. We were praying, alright—praying that it would snow until June. It didn't, but we did get sent home at lunchtime.

Miss Conley was like most adults in those days. They equated snow with crisis. My mother wouldn't say the word; she called it that "white stuff." The TV weathermen presented it as the enemy of society. If any kids my age hated snow, I sure didn't know them.

Me, I wanted to be Barry Burnell in Michigan. Barry was one of the denizens of the Corner in my neighborhood. The Corner is where the truth went to die. He would tell me fantastic stories about living in Michigan and how the snow was so deep he would jump into the snow piles outside the window of his second-floor bedroom. Years later I had the pleasure of spending a year in Ann Arbor, but I did not live the dream. Had I jumped from any second-floor window I might not have lived to share the experience. (Yes, I was in the wrong side of the state, but I'm not sure it would have been wise to jump out of second-story windows in Grand Rapids, or Marquette, for that matter.) I did experience snowfalls in Ann Arbor, incidentally not far from where James E. Church grew up. He was the Latin professor who figured out how to calculate how much water was in the Sierra Nevada snowpack, a grand reservoir for the West. Those Michigan snowfalls were memorable, but I fully understand why Church was so taken by the snows of the Rockies. I've seen summit snows there that were 12 feet deep and remnant snow in July. I once rode through a snowstorm in Nebraska that seemed to erase the entire state. I've marveled at how nonchalantly the residents of the New York Snow Belt contend with their lake-effect nemeses. I have been awed by the profoundly deep-blue hues of the sky over a Montreal snow cover—how snow can crown the peaks of New Hampshire's White Mountains well before the leaves begin to change. Yet, I have found that when it comes to snow, no venue rivals that of a childhood home, wherever that might be, and I will always associate snow with Chester.

I think of Wilson A. "Snowflake" Bentley, who had a magnificent obsession with snow that ultimately led him to change the way the world looks at snow crystals, a view later enhanced by an astrophysicist in the unlikely venue of Pasadena, California. Bentley's home was a farm in Jericho in idyllic northern Vermont during an era when families inhabited the same houses for three and four generations, where the ground stayed white all winter and the clouds would never be confused with factory smoke. I don't doubt that his powerful sense of place was a factor in the remarkable path that he chose, that what he saw in snow was a reflection of an inner life.

Since I was captured by snow before I had any awareness of conscious memories, I can't tell you why or when it might have happened. That was one of the mysteries that inspired me to undertake this project. What I did learn is that others shared the passion and had been similarly captivated. Some of them are even grown-ups. I have never grown out of mine.

The atmosphere is the life-support system. For high drama, it is unmatchable, the ultimate reality show. How tragic it is that people would prefer the hypnosis of a 55-inch TV screen to a magnificent sunset. I am grateful to snow for so many reasons. Paramount among them is how it drew my attention to the atmosphere and the wizardry of water that has the power to vanish and reappear and vanish again. I have abandoned any hopes of understanding all the riddles of the atmosphere, and I take some comfort in the fact that I have something in common with some of the best minds in science. David A. Robinson, a Rutgers University professor who is an international authority on all matters of snow, told me that his first weather memory was of Hurricane Donna ripping through North Jersey in September 1960. Yet, that is not what stayed with him: It was the snows of the subsequent winter. "Why don't I love hurricanes? Somehow, it's been the snow that's captivated me," he said. "We all try to figure out how so many of us in the weather and climate field have loved it since we were kids. . . . There must be something about it."

I make no pretense to scientific credentials; what I have to share is over-the-shoulder knowledge so generously offered by some of those who have taken on the immense challenge of trying to understand the often-maddening behavior of what scientists call a nonlinear chaotic system.

Some of the brightest and most inquisitive minds in the atmospheric and oceanographic sciences have helped me delve into the mysteries of the atmosphere. They include the one that belongs to Louis W. Uccellini, who likely knows more about winter storms than anyone on earth and now runs the National Weather Service. I am indebted to the late David Ludlum, who was the nation's premiere weather historian. His research was without parallel.

It is unfortunate that the government official recordkeeping did not begin until the latter half of the 19th century, but the journal and diary evidence, so much of it unearthed and compiled by Ludlum, provided a treasury of insights into how the first European settlers interacted with the climate on the other side of the Atlantic Ocean. I suspect that snow burrowed into the American consciousness before the nation had any say in the matter. Those bitter winters were all the more perplexing to them because the colonies were hundreds of miles closer to the equator than the places they had left. Throughout time, the new Americans not only adapted to snow, but also learned to exploit it and became quite expert at fighting it, removing it, making it, fearing it, and pining for it.

As all subjects worth pursuing, an inquiry into snow becomes a window into the natural and the metaphysical, science and history, the infinitely large and infinitely small. The views at times might be limited and frustratingly opaque, but they always are worth a look and the effort to focus the lens.

CHAPTER ONE

First Snows

Colonial Ambushes

Evoking a climate almost unimaginably enticing, Columbus portrayed the flora and fauna of the New World with effusive flourishes. In his journal entry for November 6, 1492, a time of year when harbingers of the winter chill would be evident in the Old World, he wrote of "many kinds of trees, herbs, and sweet-smelling flowers."[1] He described the "most beautiful groves of trees," as one might see deep in the month of May in Spain, and a diversity that was the "greatest wonder in the world to see."[2] The climate actually encountered by the first settlers in the North American colonies was a reasonable match for Columbus's outsized rhetoric.

In the opposite extreme.

Sagadahoc might be the oldest New World settlement that almost no one ever heard of. It was founded in the spring of 1607, just a few months after Jamestown. On the Virginia coast, Jamestown would become England's first permanent settlement in the Americas. Sagadahoc, on the Maine coast, was doomed to become a historical footnote. The Virginia and Maine colonists did share a common experience, however. They both were ambushed by shocking—and for them, puzzlingly cold—winters.

Jamestown was at a latitude similar to Sicily's, a good 1,000 miles closer to the equator than London. Yet, half the colonists died in that severe first winter. The survivors managed to endure, unlike their counterparts to the north.

Raleigh Gilbert, a distant relative of Sir Walter Raleigh, and a delegation of Britons established that colony, but it was abandoned within a year. While

Sagadahoc was hundreds of miles north of Jamestown, it still was at a latitude 500 miles south of London's. John Davies, the navigator of their ship, the *Mary and John*, wrote that the "little settlement was exposed to brutal bitter winds" and storms described as "winter severities." That first winter, Gilbert decided, would be the last. He opted for shelter on the tamer side of the Atlantic. The other settlers endorsed his superior judgment, and the colony was abandoned.[3]

What chilled the settlers were the Arctic air masses that had been prowling the continent from the frozen, forsaken regions of the planet from October to April for millennia.

For reasons unknown at the time, in England and across the channel, prevailing winds from the west softened the bite of winter. In the New World they provided the bite, howling from the mysterious, forbidding lands to the northwest of the new arrivals.

The winds had another cosmic effect. Conspiring with the strangely warm ocean current curving off the coast, the western flank of the one Columbus exploited in his New World voyages and triumphant returns to Spain, the winds ignited storms that had monstrous impacts along the Eastern Seaboard.

In the New World, the colonists would experience prodigious snowfalls the likes of which they had never witnessed, as alien to them as this strange landscape on the other side of the ocean and those stoical indigenous peoples they called "Indians." Typically, London might receive a foot of snow during an entire winter. In Massachusetts, New York, Penn's Woods, and as far south as Virginia, the colonists wrote of 2 feet, 3 feet, even 4 falling in a single storm. London, observed Puritan preacher Cotton Mather, "is never so horribly snow'd upon."[4]

When that Arctic air encountered that Atlantic current, the result could be the equivalent of an atmospheric explosion. The fallout was at the root of an American obsession that would evolve and one day lead to that most peculiar of weather-related phenomenon: the supermarket stampede.

Native Americans had snow legends that spoke to their awe and reverence for the caprice of winter. The Lenape believed that a creator designed four beings, or *mani'towuk*, to govern the winds, a leitmotif among legends of the indigenous peoples. The *mani'towuk* of the north and south engaged in a cosmic game of chance in winter, and when the winds were harsh and the snows deep, the north *mani'towuk* was prevailing. The Cherokee told of an "Ice Man" who lived in a house of ice in the far north and once extinguished a raging forest fire with a hail of rain and sleet.[5] In real life, by the time the Europeans showed up, the natives were well familiar with the New World's punishing and variable winters since their ancestors had inhabited the continent for thousands of years. What they didn't have was the Europeans' experience of suddenly transplanting themselves from the warm side to the cold side of the ocean.

The settlers' amazement at the ferocity of the New World winter storms is captured in their writings. Unlike the Native Americans and their oral traditions, the Europeans kept detailed journals; their accounts were more than legends. A lot of what they wrote about was snow.

The Selling of the New World

Captain John Smith, so instrumental in the establishment of Jamestown, had a dash of Columbus in him. Writing to Ferdinand and Isabella, Columbus foreshadowed the pages of future marketing 101 books with dramatic descriptions of an otherworldly climate. He wrote of "exceedingly fertile" lands through that flow "many broad and health-giving rivers," and lands "full of the greatest variety of trees reaching to the stars. I think they never lose their leaves."[6]

After exploring the Maine and Massachusetts coasts in 1614, hundreds of miles and a climate removed from Jamestown, he published his exuberant observations. "Smith labeled his glowing account of the bountiful fisheries, fertile soil, and varied products with a paragraph subhead: 'A proof of an excellent climate.'" Regarding Massachusetts, he wrote, "And of the four parts of the world that I have yet seen not inhabited, could I but transport a colony, I would rather live here than anywhere." Smith's writings were credited with enticing the Puritans to settle in Massachusetts Bay.

A pamphlet published in 1622 declared that in Maine, "The clime is to be so temperate, so delicate, so healthful, by reason and experience." Historian David Ludlum advised that reading the literature of the time required a healthy dose of skepticism and was about as credible as the average real estate brochure, a mix of propaganda and baloney.[7]

In 1630, William Wood, one of the first settlers of Boston, described a climate "agreeing well with the temper of our English bodies."[8]

New England indeed was closer to the equator than old England. And a popular idea circulating at the time, according to Judge Samuel Sewall, who would become famous for his role in the Salem Witch Trials, was that Boston "must be much hotter" since it was so far south of London. Sewall, one of the New World's most diligent diarists and weather observers, resoundingly debunked that concept with meticulous recordings of the severity of New England winters.[9]

The Massachusetts colonists soon discovered that American East Coast winters were radically variable and subject to prodigious snows even during the ones that overall were relatively gentle. In his journal, Governor John Winthrop described the winter of 1633–1634 as "very mild," yet it also "oft snows, and deep." In one storm, he reported, the snow accumulated two feet.[10]

Three years later, conditions would leave a deeper impression on the New World's first white settlers. "This was a very hard winter," Winthrop wrote of the 1636–1637 season. "The snow lay, from November 4 to March 23, half a yard deep." He described one snowfall as having "flakes as great as shillings."[11]

The weather got worse. "The frost was so great and continual this winter, that all the bay was frozen over," he said of the legendary winter of 1641–1642, which the "Indians" declared the coldest in 40 years, although they obviously were not abiding by the Gregorian calendar. "To the southward also the frost was as great and the snow as deep, and at Virginia, itself, the great bay was much of it frozen over, and all their great rivers."[12] Winthrop said Massachusetts was covered in three feet of snow. "This winter was the greatest snow we had, since we came into the country."[13]

Sewall wrote of how winter had chased his grandparents back to England in the 1640s. Of the winter of 1681, he chillingly wrote, "Hath been a very severe winter for snow and a constant continuance of cold weather." Of the winter 15 years later, Thomas Hutchinson, lieutenant governor of the Massachusetts Bay Colony, would describe a cold so severe, "Greater losses in trade had never been known . . . nor was there at any time . . . a greater scarcity of food."[14]

For available accounts, the snows, at once awe-inspiring and terrifying, became embedded in the consciousness of the 17th-century colonists. To some it took on mythical, mystical, supernatural proportions, and the hand of Providence was a leitmotif. The Rev. Cotton Mather saw Divine intervention in what Judge Sewall described as a "very great snow on the ground" right before Christmas in 1696. "This day, there being a violent storm arisen," Mather wrote, "I laid aside the discourse I had prepared for the congregation, and with plentiful assistances from the Lord Jesus Christ, I discoursed on the Lord Jesus Christ as a 'Refuge from the Storms of the Wrath of God.'" Mather offered a prayer for "our seafaring friends" who might be "in distress." No sooner had the prayer ended when he heard the sound of guns signaling said distress, "and Heaven sent that help unto the poor people aboard, that the vessel through extreme danger got safely in."[15]

The town clerk's official entry in the Record of Sudbury designated that season as the "terriblest winter."[16] From November 20 to April 9, a total of 31 snowfalls were reported, including 42 inches from a single storm at Cambridge. That winter was so severe that church attendance often was sparse. On a frigid January Sunday, Sewall wrote that a preacher he particularly admired, the Rev. Michael Wigglesworth, one of the Puritan poets, delivered a sermon on the topic, "Who Can Stand before His Cold." The church then closed for three weeks. When Wigglesworth returned to his pulpit, his subject was the psalm that begins, "He sends forth his word and thaws them."[17]

A thaw began the next day.

"Eyes Glazed Over with Ice"

By the turn of the 18th century, the population was growing rapidly in the colonies, quadrupling from 1650 to 1700, when the total reached about 250,000. More people meant more needs and more impacts from snow. With the consent of the mother country, an intercolonial postal system was established linking Massachusetts, New Hampshire, New York, and Pennsylvania. It was no match for the winter of 1704–1705, however. In what was believed to be the first public account of snow-related road closings in the New World, a succession of profoundly heavy snows from Philadelphia to New England knocked out the nascent mail service for six weeks. Drifting to impossible depths, the snows closed all the postal roads. Philadelphian Isaac Norris wrote, "We had the deepest snow this winter, that has been known by longest English-liver here. (*Liver* would be inhabitant, not organ.) No traveling; all avenues shut; the post has not gone these six weeks."[18]

The winter of 1704–1705 and the following one were among the 25 in the 18th century that Ludlum classified as either "very severe" or "severe." An Arctic outbreak that arrived right at the end of 1705 was blamed for the fatalities of 132 crew members who froze to death after their ship was grounded by gale-force winds not far from New York.[19]

Testifying to the extreme nature of those winters is the fact that the winter of 1716–1717, the year of the so-called Great Snow, didn't make the cuts on either of the two lists. In a letter to the Royal Society, Cotton Mather spoke of a February 20 storm that year "so violent as to make all communication between neighbors everywhere to cease." Four days later, a Sunday, "Another snow came which almost buried the memory of the former, with a storm so famous that Heaven laid an Interdict on the religious assemblies throughout the country."[20]

Mather frequently corresponded with the Royal Society to share his New World scientific observations, and Thoreau considered his account of that winter as the masterwork of the "Keeper of the Puritan Conscience." When the snows melted, Mather said that cattle "were found standing dead on their legs" and "others had their eyes glazed over with ice." Of 100 sheep pulled from 16-foot snow piles, he wrote, 98 were dead, and two "kept themselves alive by eating the wool of their dead companions." Orchards were destroyed, "for the snow freezing to a crust, as high as the boughs of trees, anon split them to pieces."[21]

Mather viewed the wintry assault as indicative of a "time of much rebuke from Heaven upon us. . . . Such storms and heaps of snow, visit us in the approach of spring, as were hardly ever known in the depth of winter." He advocated a "day of humiliations and supplication."[22]

According to the government's official account, the February 20 and 24 storms weren't even part of the Great Snow, which actually consisted of four snowfalls from February 27 to March 7. As much as five feet were observed in Boston, with drifts covering second-story windowpanes.

In the first four decades of the 18th century, the colonies experienced 10 winters on the Ludlum severe lists. Then came the "Hard Winter" of 1740–1741. "The elements have been armed with such piercing cold and suffocating snows," the Rev. James MacSparran told his congregation in Narragansett at the end of that winter, "as if God intended the air that he gave us to live and breathe in, should become the instrument to execute his vengeance on us for our ingratitude to his goodness."[23]

Ludlum said he never found evidence of a winter before or after to match that one for persistent cold and heavy snows. Yukon air masses that poured across the Northeast after a succession of storms locked waterways in ice from the Mid-Atlantic to New England. Boston Harbor was frozen solid for 30 days. A piece of Long Island was joined to the mainland by a highway of ice, and shipping at Philadelphia was suspended from December 27 to March 27, as icing closed off the Delaware River. Snow drifts choked highways throughout the Philadelphia-to-Boston population corridor.[24]

From 1700 to 1750, the population of the colonies quadrupled yet again, to almost 1.1 million, and since so much of that growth was from Virginia to Massachusetts, the impacts of snow were becoming more significant, and the accounts more plentiful.

Perhaps the most well-documented snowfall of the colonial era was the "Washington and Jefferson Snowstorm," so designated because descriptions of it appear in both men's diaries.

It was the "deepest snow we have ever seen," Thomas Jefferson wrote. In Albemarle, near Monticello, "it was about three f. deep." The storm was an ordeal for Jefferson and his new bride, who were returning from their honeymoon on the afternoon of January 26, 1772, when the snowfall became so heavy that they had to abandon their carriage and struggle through the intensifying snows on horseback over a mountain trail for eight miles. "They arrived late at night," Jefferson's daughter wrote, "the fires all out and the servants retired to their houses for the night. The horrible dreariness of such a house, at the end of such a journey, I have often heard both relate."[25]

The snow evidently started later in Mount Vernon, which would have been a typical precipitation pattern for an Eastern snowstorm. George Washington reported that it had begun during the night and that five or six inches had accumulated by morning. It continued throughout the day and into the next, and the future first president complained that he was confined to a white prison, "the snow being up to the breast of a tall horse everywhere." Several inches were recorded as far south as Winston-Salem, North Carolina. Two years later, snow would fall in Florida, although colonial historian John Lee Williams, perhaps concerned about the future Sunshine State's image, used the euphemism "white rain."[26]

History's Most Famous Wintry Mix

Three of the Ludlum 18th-century severe winters coincided with the American Revolution. Malnutrition and exposure were blamed for contributing to 2,500 deaths at Valley Forge during the encampment in the winter of 1777–1778, but that one also failed to make the Ludlum lists. The preceding winter did, and its signature moment was one of the most famous winter storms in history. A powerful nor'easter commenced affecting the Mid-Atlantic region on Christmas 1776. Jefferson reported measuring 21 inches of snow at Monticello. About 250 miles to the northeast, in the areas around Philadelphia, the snow totals were about half that, but conditions were made all the more punishing by the sleet and freezing rain that held down accumulations. That is a common occurrence during storms in areas closer to the ocean—the Atlantic is only 60 miles from Philadelphia—as winds from the east import warmer air aloft that causes a precipitation changeover, while the cold at the surface is more tenacious. The result is sleet, which is melted snow that refreezes on its way to the ground, or freezing rain, which is rain that freezes on contact.

Those icy conditions have been forever frozen in American lore. On Christmas night, Washington executed his legendary crossing of the Delaware from north of Philadelphia. Washington's troops marched nine miles in the punishing mixture of snow and falling ice to Trenton, famously engaging and defeating the Hessian soldiers allied with the British.

Ultimately, the American victory in the Revolution would bring profound change to the New World. The loosely allied 13 British colonies would become the United States of America. Some argued that the climate was changing, both the political and the natural, a subject that would come up again.

One thing didn't change: It continued to snow.

CHAPTER TWO

Snowmen

Giants of Snow Research

The settlements of the New World coincided with the predawn era of snow-flake research in the Old World. On a practical level the ongoing pursuit of that science holds immense potential value to transportation interests and highway departments that try to keep wheels and wings in motion, and to meteorologists, who have long struggled with accumulation forecasts, often to the vexation of an impatient public.

An unlikely modern figure in this pursuit of the enigma of the snow crystal works in the unlikeliest of venues. Kenneth G. Libbrecht is a powerhouse in the field of astrophysics, an award-winning professor at Caltech, one of America's leading collection points for brainiacs. He is among the scientists involved in the LIGO—Laser Interferometer Gravitational-Wave Observatory—project, which is in a grand hunt for Einstein's ripples in space-time. Libbrecht is also in a grand pursuit of the enigmas of something that falls from the sky on Planet Earth: the snowflake, what he sees as one of the more fascinating manifestations of the mysteries of the universe. Never mind that Caltech is located in Pasadena in deep Southern California, where the flaki-ness ordinarily has nothing to do with snow, and, yes, Libbrecht is familiar with the stereotyping and doesn't take it personally.

It is estimated that about half the world's population has not experienced snow. Although the city is several hundred feet above sea level, until Febru-ary 21, 2019,[1] it was safe to assume that most of the residents of Pasadena likely would have fit into that category. That day flakes were sighted, although they vanished as magically as they had appeared; at the nearest official measuring station, located at Los Angeles International Airport, the snow total for the day was 0.0. The last time snow in Pasadena had registered anything on the centimeter side of the ruler—and in that case, 1.27 centimeters, or 0.5 inches—was on January 21, 1962, when Richard Nixon was running for governor of California.[2] When I visited Libbrecht and his laboratory, a splendidly chaotic exhibit of Styrofoam, FlexFix silver tape, spaghetti bowls of wires, and assorted other paraphernalia, it was a cool-ish day for early February in Pasadena—as in temperatures in the upper 50s. That didn't bother him; his lament was that the endless Pasadena summer was only days away.

So why is he so interested in snowflakes? It has nothing to do with the challenges of accumulation forecasts or promoting the efficacy of freeing the nation's transportation network from the bondage of snow and ice.

"I feel that with over 6 billion people on the planet," he has said, "surely a few of us can be spared to ponder the subtle mysteries of snowflakes."[3] He is not sure where it might be leading, and that's almost beside the point. Oscar Wilde once said that all art is useless; Libbrecht said a similar case could be made for much of science. He says he is often asked what is the practical application of his snow research; he counters that no one asks the same question about astrophysics. He notes that scientific history is full of examples of interest-driven, motive-less research that ultimately had quite significant unforeseen impacts. He has cited Michael Faraday, the 19th-century physicist, who was asked what could possibly result from his discovery that magnetic fields could produce electric currents. Mathematical research conducted in the mid-19th century formed the bases of search-engine algorithms. Take liquid crystals. Chemists became intrigued with them in the 1920s because of the peculiar visual effects they created. No one conceived that they would one day be ideal for video-game displays.[4]

Libbrecht said he likes to tell his students, "Enjoy the process." It's okay to learn about something simply because it piques your interest. When we met, he was writing a book about the science of snow crystals, confident that no one would read it. So what? He has pursued snow crystals because they *are*. Among contemporaries, he has opted for a detour to follow a path less traveled by, visiting places he never had anticipated his research would take him. In doing so, he is aware that he has followed in some giant footsteps.[5]

A Six-Sided Enigma

While Europeans were settling in the New World, in the Old World, Johannes Kepler, Rene Descartes, Robert Hooke, and other intellectual titans were redefining the frontiers of science and mathematics. Kepler's laws of planetary motion, Descartes's fusing of geometry and algebra, and Hooke's hypothesis about the origins of color would alter the ways humans perceived the universe. The three had unique, disparate, and expansive interests. This is an age of specialization, and while Libbrecht agrees that is true in physics—and also especially true in meteorology and oceanography—once upon a time it was acceptable for scientists to care about more than one thing.

Kepler, Descartes, and Hooke did share at least one interest: the snowflake. Nature is the ultimate magician, its mastery ever evident in the visible and invisible world of water, and as Libbrecht would centuries later, Hooke, Kepler, and Descartes all indulged in the spell of the crystals. The written record of the ongoing pursuit of the scientific mysteries and the cross-examination of the snowflake more or less begins with them. Their analyses and ruminations were at turns learned, witty, profound, and insightful. From a personal perspective, I was delighted to learn that the snowflake had found its way into the prodigious minds of people with such diverse interests, the way it can find its way into the architectural niche of a cathedral. It also made me less self-conscious about pursuing my own interest. Snow is a great equalizer.

Wilson Alwyn Bentley had about as much in common with Kepler, Descartes, and Hooke as Ken Libbrecht and this author, which is to say less than not much. He was two centuries removed from the Scientific Revolution and a homeschooled Vermont farmer. Yet, with no formal training whatsoever, he changed the way humans perceived the snowflake—more properly, the snow crystal. This diminutive man became a towering figure in the history of snow science. He was the one who first exposed the inner lives of these "crystal masterpieces" that sprouted magically upon the tiniest of particles from volcanoes, from the reproductive fallout of plant life and interstellar dust. With a primitive but inventive contraption he crystallized what had captured Kepler, Descartes, and Hooke.

The Invisible World

Bentley's entree into the secret cosmos of the snowflake was the invention that revealed the architecture and inner life of the invisible world: the microscope. This deserves more than a few words about its role in revealing the artistry of the snow crystal.

If science had a Hall of Fame, the inventor of that ingenious device would have an exalted place. Unfortunately, no one is quite certain who that person or persons might have been or precisely when the *eureka!* moment occurred. Magnification is an ancient concept; with the invention of glass, the Romans were well familiar with the concept. But the "compound microscope" as we know it wasn't developed until deep into the Renaissance. A pair of Dutch opticians, Zacharias and Hans Jansen, are credited with assembling the first one.[6] In the late 16th century, they discovered that by stacking several lenses inside a tube, the object at the end appeared dramatically larger. Galileo, who might have been the first to use the term "microscope," was quoted as remarking that "with this long tube" he had seen "flies which look as big as a lamb."[7]

Kepler saw the potential, and both he and Galileo are credited with important insights that led to the perfecting of the concept[8]; however, both Kepler and Galileo would claim their exalted places in scientific history by focusing on the larger universe rather than the heretofore unimaginably small. As for snowflakes, if Galileo or Kepler used magnification to investigate the riddles of the crystals, they left no record of it.

Kepler's attention was first drawn to them providentially, he wrote in his whimsical 1611 treatise, *De Nive Sexangula*. While walking across a bridge in Prague, "specks of snow" fell upon his coat, "by happy chance."

Kepler held that the key to a flake's career was embedded in the fundamental character of water vapor. "There is thus a formative faculty in the body of the earth, and its vehicle is vapor, just as breath is the vehicle of the human soul." He offered several hypotheses for the fact that all the crystals were hexagonal, concluding that it was not related to a survival strategy or anything that could be identified clearly as practical. Nothing in the shape suggested that it would ensure a crystal's safe landing, for example. The snowflake, he suggested, might well exist for the pleasure it might bring to the observer.

The hexagonal consistency, he opined, "cannot be chance. Why always six?" Straddling the Age of Faith and the Scientific Age, Kepler postulated a supernatural explanation: "Supreme reason . . . I cannot believe that this ordered shape is present by chance."[9]

The hexagonal repetitions constituted an ancient source of curiosity. Chinese philosophers had remarked on the peculiar six-sidedness as early as 135 BC. Chinese scholar Han Yin observed that snow crystals are "always six pointed." In the 6th century, poet Hsaio Tun wrote, "The white snowflakes show forth their six-petaled flowers."[10]

More than a thousand years later, Descartes was struck similarly by the symmetry of the crystals, but he also saw in the shapes of the crystals evidence of dramatic and chaotic careers. Libbrecht pointed out that Descartes was a physicist. Descartes was reveling in the "euphoria" of his "discovery of the scientific method," observed British physicist Francis Charles Frank.

"Descartes could persuade himself of his ability to explain anything." Yet, that "blended strangely, almost uniquely, with his capacity for exact perceptive observation." Remarking on what he witnessed in snowfalls in 1635, Descartes wrote, "I only had difficulty to imagine what could have formed and made so exactly symmetrical these six teeth around each grain in the midst of free air and during the agitation of a very strong wind." He then deduced that the wind was a promoter, not a saboteur, of a crystal's symmetry.

> This wind had easily been able to carry some of these grains to the bottom or to the top of some cloud, and hold them there, because they were rather small; and that there they were obliged to arrange themselves in such a way that each was surrounded by six others in the same plane, following the ordinary order of nature.[11]

As were Kepler and Galileo, Descartes was familiar with microscopes. Yet, as with his predecessors, if he ever considered using them to examine snowflakes or any other objects, he left no record. It was not until a generation after Descartes's ruminations that such a record was produced.

Just as the microscope introduced the invisible world to the scientific community, Robert Hooke drew the world's attention to the microscope. His astonishing *Micrographia*, the first major attempt to publish images of this magnified universe, became the first science book to make the best-seller list. Diarist Samuel Pepys purchased a copy of *Micrographia*, and when he got around to reading it several days later, he literally couldn't put it down. He pored through it until 2 a.m. and declared it "quite remarkable" and the "most ingenious book that I ever read in my life."[12]

"I made use of microscopes and some other glasses and instruments that improve the sense," Hooke wrote, "both in the surveying the already-visible world and for the discovery of many others hitherto unknown." He declared exuberantly, "By this the Earth itself, which lies so near us, under our feet, shows quite a new thing to us, and in every little particle of its matter; we now behold almost as great a variety of Creatures, as we were able before to reckon up in the whole universe itself."[13]

Hooke was a prominent member of the Royal Society and had worked as a lab assistant to Robert Boyle, and is said to have built the vacuum tubes that helped Boyle discover the gas laws. A philosopher and an architect, Hooke had eclectic interests, as did other intellectuals of his era. Due in large measure to *Micrographia*, however, he would be most remembered for his work with the microscope. In an effort to share with readers what he, himself, saw, Hooke applied a talent for drawing that he had developed in his youth.

The book, published in 1664, was a graphic sensation, and not just in the opinion of Pepys, with engravings and fold-out pages of his drawings of insects. Hooke had a reputation for being irascible and combative (then again, so did Beethoven), but he obviously had a highly cultivated sense of wonder that added fresh perspectives on ordinary objects.

Through the power of the microscope, he found, insects otherwise unexceptional to the naked eye—the flea, the louse—were transformed into alternately monstrous, delicate, and complicated beings. When Hooke examined a fragment of cork, he saw structures that evoked honeycomb cells or the monastic cells in which the monks slept, and he is believed to have been the first to use the term "cell" in the biological sense.

Hooke was determined to capture his subjects' architectures with as much resolution as possible. One particularly difficult subject was an ant that refused to play the cooperative model. Hooke acted. "I gave it a Gill of Brandy, or Spirit of Wine," he said, "which after a while e'en knock'd him down dead drunk, so that he became moveless."[14]

He also attempted to capture images of an even more elusive object: the snowflake. His drawings represented the most ambitious attempt yet to introduce the public to their appearances in magnified form. Hooke admired an "infinite variety of curiously figured snow" and included several examples in his book. A few of them look more like fish bones than snow crystals, but most do at least capture a variety of hexagonal shapes.[15]

While London was less generous with snow samples than the colonies that constituted the "New England Confederation," Hooke was living in the heart of the Little Ice Age, an era when temperatures were significantly lower than they are today.[16] For a variety of reasons, England did not experience snowstorms of the New World's magnitude, but the London winters almost certainly were snowier than they have been in the 21st century and the snow cover more tenacious after it had landed. Thus, Hooke would have had plentiful subject matter.

The bigger challenge, as Wilson Bentley later discovered, much to his frustration, would have been replicating the intricate details of such an evanescent subject.

Hooke, an astute observer, said of the crystals, "The bigger they were magnified, the more irregularities appeared in them." He hypothesized that before its descent a crystal had existed in a Platonic state but that its perfection was compromised by the "thawing and breaking of the fall." *Micrographia* does not discuss how he transferred the flakes to his slides, and it is possible that melting and evaporation had something to do with the deterioration of form; however, his insight that a flake would undergo changes on its earthward journey represents an important clue about the changes that snow underwent on a trip that could last for three hours.

Hooke also was struck by the same aspect of the flakes that had drawn the attention of Chinese philosophers and Kepler and Descartes: the signature geometry. "They were always branched out with six principal branches all of equal length, shape, and make from the center."[17]

For whatever reasons, the curiosity exhibited by Kepler, Descartes, and Hooke about the origins and artistry of snow crystals was absent in the 18th century. That is perhaps surprising given the eclectic interests of New World Enlightenment figures Thomas Jefferson and Benjamin Franklin, both diligent weather observers. The exploration of the snowflake's secret, however, would take a quantum leap in the 19th century.

A Farmer's Dream; a Father's Rebuke

Wilson Bentley was a slightly built teenager who worked on the family farm in the heart of Vermont snow country. He was the son of a former schoolteacher. He also was his mother's prized pupil, and she educated him at the Bentley homestead in Jericho. If "Willy" Bentley ever did spend time in a conventional classroom, it wasn't much, recalled Sue Richardson, his great grandniece, who grew up in the Bentley household at a time when it was common for four generations to occupy the same dwelling.[18] Willy, achingly shy, passed whatever free time he had devouring every book he encountered. On February 9, 1880, his 15th birthday, his mother gave him a life-changing gift, a microscope she had used in her teaching days. Willy became addicted to the magic of magnification. He began looking at anything he could place on a slide. Biographer Duncan "Dune" Blanchard, a Ph.D. atmospheric scientist at MIT, declared him the world's first cloud physicist.

"When the other boys of my age were playing with popguns and slingshots, I was absorbed in studying things under this microscope: drops of water, tiny fragments of stone, a feather dropped from a bird's wing, a delicately veined petal from some flower," Bentley wrote. Bentley authored an authoritative treatise on raindrops and another on human smiles—seriously.[19]

But what captured his attention more than any other object? "Always, from the very beginning, it was snowflakes that fascinated me most," he reflected. "The farm folks, up in this north country, dread the winter; but I was supremely happy, from the day of the first snowfall—which usually came in November—until the last one, which sometimes came as late as May."[20]

Jericho is about 15 miles east of Vermont's largest city, Burlington. Chilled by stinging winds off Lake Champlain, the Burlington winter has a hard-earned reputation as long, frigid, and snowy. In the highlands of Jericho, winters aren't quite like Burlington's: They are longer, more frigid, and snowier. Thus, Jericho was an ideal venue for Bentley to indulge his fascination. That, he did.

Bentley was an accomplished musician, and Dune Blanchard tells the story of the night that Bentley was the pianist at a farmhouse square dance, a popular seasonal ritual during the enduring Jericho winters. Right before the dance ended, several boys went outside and removed the wheels from his buggy and replaced the order, mounting the larger wheels on the front and the smaller ones on the rear. "He went home like that and drove the buggy several days before he noticed it," a witness told Blanchard. "I don't know what you'd call that."[21] Call that anything but absentminded, which, along with the likes of "near miss," is one of the more egregious misfires of the language. Bentley was very much present-minded. He deconstructed the snowflake with a laser focus.

For all the raw material he had to work with, finding just the right specimens for scrutiny was his first challenge. Historically, the term "snowflake" has been used interchangeably with "snow crystal." The latter, which is scientifically correct, refers to an individual six-sided crystal. A snowflake can be a conglomeration of hundreds of crystals and fragments. Bentley's research affirmed that Descartes was correct: The earthward journeys of snow crystals could be the antithesis of the tranquility ultimately affected by a generous snow cover. Some crystals smash into one another, breaking into bits, or latch on to others that collect crystalline material. To the observer at the surface, these conglomerates might appear as a single flake, but Bentley knew the distinction and always was careful in his technical writing to use the term "snow crystal." When he sought his specimens for observation, he looked for the loner crystals that had survived the trip or those that had broken off from the crowd when a conglomerated flake smashed into a surface.

Once Bentley identified and captured his subjects, the obstacles became more formidable. He set up his "laboratory" in an unheated woodshed that was more like a walk-in freezer. The average January temperature in Jericho was in the upper teens, or about 50 degrees below room temperature, but few things in the known universe are more evanescent than the snow crystal, and nothing would accelerate the vanishing process more than heat. While examining the flakes, he had to hold his warm breath for as long as he could.[22]

Among the "necessary accessories" for crystal acquisition that Bentley itemized were "a pair of thick mittens; . . . a feather, and a turkey wing or similar duster," and a one-foot-square blackboard "with stiff wire or metal handles at the ends, so that the hands will not touch and warm it."

He caught the falling snow on the blackboard and scanned it with the naked eye or through a magnifying glass to identify the candidates. He used the feather duster to keep his dark staging platform clean until "two or more promising specimens alight upon it," at which point he would whisk the board into the unheated shed, racing against the corruption or outright extinction of his subjects. Said Bentley, "The utmost haste must be used."

He transferred his "promising specimens" from the blackboard to the microscope slide with a sharply pointed piece of wood that he placed in contact with a crystal until it adhered to it. Typically, several crystals would occupy a single slide, each receiving an instant's glance. Multiple crystals exuded a cooling that had the advantage of supplying a refrigerant effect throughout the slide. Bentley held his breath while he examined his slide. With even a trace of exhalation on them, the subjects were lost forever.[23]

Bentley would exhale away from the microscope and then attempt to draw what he had seen.[24] Melting wasn't the only threat to crystals' lives, however. They were quickly compromised and eroded by evaporation, particularly at the ultra-delicate tips. It didn't matter if the temperature was below zero. The chameleonic qualities of water are indefatigable. By Blanchard's estimate, a snow crystal, even in that frigid room, would have survived only a matter of minutes. Imagine attempting a detailed drawing of a model who could pose for only a few minutes. This was what Bentley was up against. For three years he tried his hand at illustrating the crystals he had selected so carefully, but knew he had to come up with another way to do justice to these natural masterworks. Fortunately, he was about to get a major assist from an emerging technology.

A Picture of Tenacity

In the early 1880s, photography was a nascent and developing pursuit. The Bentleys had a family relative, Henry Seeley, who operated a portrait studio in Connecticut, Richardson recalled.[25] Seeley often came back to his native Jericho for visits and conversed with Willy frequently. "I got him his first camera," Seeley told Blanchard, "and showed him the making of plates, and whenever there was a new book published on photography, I used to send it to him." Bentley, he recalled "was a genius," and a quick study. The student became a teacher. "It was not many years before he could give me information."[26] How Bentley learned about the art of microphotography is unclear. Since he read everything he could get his hands on, it is possible he came across technical treatises about a process developed in Europe coupling the powers of the camera and microscope. By the mid-19th century, scientists and physicians were using photographs of enlarged images for serious research purposes. Evidently no one had considered blending the microscope and the camera to secure photographic images of snow crystals.

Given his prodigious imagination and creativity, it is also possible that Bentley intuited that he could fashion such a contraption. However he learned of the possibilities, he confronted a fundamental barrier. He would need a camera, a better microscope, and for both he needed something neither he nor his family had much of: money.

Bentley was 17, and with his duties around the farm, from milking cows to working the potato fields, he had no opportunity to supplement his earnings. A camera and a new microscope, devices that had absolutely nothing to do with the demands of agriculture in an age predating labor-saving machinery, were extravagances that a Vermont farm family hardly could afford. His father was resolutely opposed to the purchases, so Wilson Bentley did what so many offspring have done through the ages: He appealed to the seat of power in the Bentley household, his mother, Fanny Colton Bentley. And as mothers have done through the ages, she prevailed.

His father agreed to part with 100 precious dollars—roughly $2,500 in today's currency—for something he viewed as useless, although ultimately, he parted with the money unresentfully.[27] The father didn't realize that he was making an immense contribution to the history of snow science. In fact, he never would. "You can imagine, or perhaps you cannot, unless you know what the average farmer is like, how my father hated to spend all that money on what seemed to him a boy's ridiculous whim. . . . He never came to believe it was worthwhile," said Wilson.[28]

How the Bentleys managed to come up with that $100 was long a mystery, finally solved by Sue Richardson, an avid genealogist. She discovered that after Fanny Bentley's parents died, they left her a bequest that happened to include $100. That was the money Willy used for his equipment.[29]

Bentley acquired a bellows camera and the least expensive Bausch & Lomb microscope available. His plan was ingenious. The microscope was equipped with a joint so that the upper part could pivot horizontally. The accordion-like bellows provided a light-less corridor. Inside the unheated room he mounted his makeshift apparatus atop a three-foot-high table with a two-square-foot surface. To his frustration, Bentley, who had short arms, discovered that he was unable to reach the microscope's focusing knob from the rear of the camera. With the rapidly changing and melting crystals, if he wanted to capture the images in a pure state, he could not spare the precious time required to shuttle back and forth from the knob. So with string and pieces of wood, he rigged up a pulley device that allowed him to focus while staying put.

Bentley removed the conventional camera lens, and his "shutter" was a black card placed over the objective. To take his pictures, he removed the makeshift shutter and replaced it after the exposure period, which varied from eight seconds to almost two minutes. And the results? Lamented Bentley, "I failed over and over again!" A farm was no place to seek technical assistance. Bentley understandably was anxious. "Here I was with this expensive apparatus that had been given to me so reluctantly. I had been sure I could do wonderful things with it. The winter slipped away, and I was almost heartbroken." The winter of 1883–1884 passed without a single successful photograph.[30]

After that particularly frustrating first winter, Wilson Bentley, himself, worried that his father was onto something, that the entire project was an expensive waste of time and precious money. Bentley defaulted to a trait that served him well: persistence. The following winter, he made some crucial adjustments in lighting[31] and increased the exposure times. It is uncertain whether he had consulted with Henry Seeley or anyone else, but on January 15, 1885, as five inches of snow lay outside his woodshed, it was clear that Bentley had solved his problems. At age 19, on that day he produced the very first photomicrograph of a snow crystal. "I knew then that what I dreamed of doing was possible," he said. "It was the greatest moment of my life!"[32]

Bentley's fame wasn't exactly instant. He lived off his share of the modest income from the family farm, which he had to supplement by giving music lessons. All the while he continued his microphotography of snow crystals. He was articulate and eventually wrote about his findings and hypotheses prolifically, sometimes poetically, and with erudition. That writing facility, too, would come later, however.

Although Bentley knew he had accomplished something startling, his achievement evidently wasn't celebrated, even in his household. That would come later, Richardson said.[33] As for sharing it with the world, in Blanchard's view Bentley clearly was self-conscious about his lack of academic credentials and believed that he could add nothing to the knowledge base of the scientific community. He also was insecure about his writing ability. Bentley allowed that he once submitted an article to a New York paper but that it had been rejected. "I can't write well, you know," he once told an interviewer.[34]

If Bentley was discouraged about his ability to communicate his findings, it didn't stop him from continuing to take his amazing photographs, and by March 1898, he had a portfolio of 400 of them. That year his reputation took a giant step outside of Jericho. He made his first big sale, although "sale" might not be the appropriate word. The Harvard Mineralogical Museum accepted his entire collection—the arrangement that did not fatten his bank account.

According to an article in the *Proceedings of the American Academy of Arts and Sciences* dated April 13, 1898, Bentley "sold" the prints for a "nominal cost." What Bentley received in return, however, transcended money. The article had high praise for Bentley's meticulous presentation, noting the collection's "scientific value is enhanced by his notes," which included extensive observations about the meteorological conditions at the times the crystals were photographed. It said the images constituted a "monument to the patience, skill, and enthusiasm of the maker."[35]

Perhaps emboldened by his success with the Harvard museum, Bentley intensified his efforts to "share my treasures" with the rest of the world. To that end, he made the 16-mile trek west to Burlington, armed with some of his crystal images.

He took his treasures to the University of Vermont to show them to Professor Henry Perkins. In Perkins, Bentley found a kindred spirit, a throwback to the Scientific Age in that he was a man of eclectic interests who had written several books on botany and geology. It is possible that Bentley came upon his name when devouring encyclopedias at his home; Perkins had written articles for *Encyclopedia Britannica* and others.[36]

More likely, said Richardson, Bentley had not set out specifically to meet with Perkins but wandered about the campus seeking advice on just who might be interested in his photographs[37]; however, he ended up visiting Perkins, and he couldn't have made a better choice. Examining Bentley's images, Perkins recognized that he was in the presence of something special. He was amazed at the quality and told Bentley he absolutely must write about them. "I tried," said Bentley. By his account, he failed, and he had Perkins author a paper for him. Perkins's account of that paper's production was quite different. In the footnote to the paper bearing both their names, he gave almost a lion's share of the credit to Bentley, saying the important aspects of the article were the result of Bentley's "untiring and enthusiastic study of snow crystals."[38] As for the writing, the paper bears clear evidence of the hand of Bentley.

The remarkably perceptive article was published in the May 1898 issue of *Popular Science Monthly*. Without specifically mentioning Robert Hooke or Bentley's own previous efforts, the authors noted that heretofore, "skillfully executed drawings" were the only means of examining a variety of snowflakes. No drawing, however, could capture the intimate details of a flake's interior life. By contrast, as Perkins must have realized, Bentley's revolutionary photomicrographs penetrated the crystal's deeper secrets. As for the writing, certain passages paralleled Bentley's signature prose style that would appear later in solo articles.

The phrase "no two snowflakes are alike" is often attributed to Bentley. Libbrecht quite accurately opined that just about everyone at some point has heard the phrase and virtually no one can remember where he or she first was imprinted with that axiom. Maybe it is written on the wind. But what does it mean, exactly? By simple observation, snowflakes do appear to be very much "alike," and perhaps the phrase "no two snowflakes are alike" was our first introduction to the concept that things aren't quite what they seem. While most people understand the concept expressed by the phrase, the wording, itself, is imprecise, and it is not how the concept was expressed in the Bentley–Perkins paper. "So varied are these figures that, although it is not difficult to find two or more crystals which are nearly if not quite the same in outline, it is almost impossible to find two which correspond exactly in their interior figures," they wrote. The article exhibited a precocious understanding of snowflake morphology that would have made Descartes proud, if not envious. It hypothesized that crystals that originated in higher, colder clouds tended to be finer and "less branched" than those from the warmer environments of lower clouds; however, the crystals from on high, "as they fall through layer after layer of clouds, each layer subjecting them to its own special conditions, they may be greatly modified, and by the time they reach the earth they may closely resemble the crystals from lower clouds." Nevertheless, "they can usually be distinguished from them by an examination of the internal structure, as well as by, in some cases, their general form."

A careful examination of the structure

> not only reveals new and far greater elegance of form than the simple outlines exhibit, but by means of these wonderfully delicate and exquisite figures much may be learned of the history of each crystal, and the changes through which it has passed in its journey through cloudland. Was ever life history written in more dainty hieroglyphics.[39]

These observations bear Bentley's stylistic imprint as plainly as crystals bear the imprints of atmospheric conditions encountered on their earthward journeys; Ken Libbrecht, who has read and written more scientific-journal prose than most living human beings, affirmed that the term "cloudland" is not the stuff of academic journals.

The article's publication was a breakthrough for Bentley. He would go on to publish 50 popular articles and 11 technical papers. In 1920 and 1921, his work was the subject of articles that appeared in the *New York Tribune*, *Boston Globe*, and *Philadelphia Public Ledger*, the predecessor of my employer. In a piece carried in the *Christian Herald*, Bentley described snowflakes as "gems from God's own laboratory." He wrote,

> How wonderful, how very complex must be the processes going on up above at such times. Could we but boat upward with these wonderful water molecules and see the snow crystals form, atom by atom, spangle by spangle, how absorbingly interesting would be the experience, and what wondrous secrets and hidden processes we might witness.[40]

Bentley meticulously cataloged his microphotographs, from number 1, on January 15, 1885, to number 5,381, on March 1, 1931, recounts biographer Blanchard. Just a few months before he died of pneumonia, two days before Christmas 1931, McGraw-Hill published *Snow Crystals*, showcasing almost 2,500 of Bentley's photographs. As testimony to his growing stature in the meteorological community, his collaborator was William Humphreys, chief physicist at the U.S. Weather Bureau, who helped raise money for the project. Humphreys, then president of the American Meteorological Society (AMS), had advocated passionately for preserving Bentley's work. In an eloquent speech in front of AMS members in May 1930, he argued for the urgency of saving what he called "meteorology's frozen assets." He said that while many of Bentley's photographs were available for public view, countless more in Bentley's custody could be lost by an act of nature. With the publication of *Snow Crystals*, the work of this "rare and kindly genius," Humphreys wrote in the introductory text, "would be made liquid—readily available to the scientist, the artist, the layman."[41] I would say that Mr. Humphreys, himself, had a writerly touch.

From Jericho to the World

The publication of *Snow Crystals* marked a watershed in the history of snow-crystal research. The book came into the hands of an admirer more than 6,000 miles away, brilliant Japanese nuclear physicist Koichiro Nakaya, one of Libbrecht's heroes. Inspired by Bentley's work, in 1932 Nakaya began his own investigation of the snow crystal, adding new layers of insight atop the knowledge accumulated by the Vermont farmer.

Being a physicist, Nakaya had difficulty finding work in Tokyo, Libbrecht said, so he ended up accepting a job as a professor of physics at Hokkaido University in Japan's northernmost island. Hokkaido was an elite university—a Japanese MIT or Caltech.[42] When he accepted the position the department was poorly funded, and it had no nuclear facilities. What the campus did have was snow. The university was located in the city of Sapporo, which has an annual snowfall of almost 200 inches. Nakaya, the nuclear physicist, took to studying snow crystals. He mined the nearby mountains for specimens, and following Bentley's model he used a microscope-camera apparatus to assemble 3,000 images. Nakaya then took a monumental step.

It was an era in which Japanese scientists were racing ahead of the Western world in atmospheric research. In the 1920s, Wasaburo Ooishi discovered high-altitude winds that became known as the "jet stream." Although he published his findings in Esperanto, the universal language, Ooishi's research was ignored outside of Japan, to the peril of U.S. and German pilots who encountered those winds in World War II as their planes were reaching new heights, costing the planes valuable fuel, blowing bombs off target, and endangering the crews.[43] A decade after Ooishi's landmark paper, Nakaya's inventiveness would lead to a cosmic development in unlocking the interior life and times of the snowflake, and become a model for Ken Libbrecht.

For all the snow at his disposal, for scientific analysis Nakaya wanted a measure of control over his subjects that nature would never bestow. Physicists are like that, Libbrecht told me. As Libbrecht recounts, Nakaya had developed a specialty in wintry precipitation. His reputation for expertise on the subject was so recognized that he was able to lobby successfully for funding from the imperial government for what became the Low Temperature Science Laboratory, housed in a wood-paneled refrigerated room. His plan was revolutionary: He would create his own snow crystals by cooling and condensing water vapor developed in a hollow glass tube. As Libbrecht would become aware, however, producing snow crystals in a laboratory is a tad more complicated than making ice cubes. Those tiny particles of volcanic or cosmic dust in the atmosphere, the sine qua non nuclei upon which snow crystals congregate, were absent in Nakaya's laboratory. Nakaya tried various substitutes: the likes of ultra-fine string filaments of silk and wires. and spider-web threads. But instead of being gathering points for crystals, the materials attracted frost particles that more resembled frozen worms than stellar dendrites. Ultimately, he found the solution: the hair of a hare. He discovered that the natural oils of a rabbit hair repelled the homely frost crystals and was the ideal meeting point for the crystals. On March 12, 1936, he produced the world's first man-made snowflake.[44]

Nakaya could now provide some answers to the questions raised by Kepler, Descartes, Hooke, and Bentley, supplanting their speculations with remarkable detail. In the laboratory, he was able to produce crystals in almost every condition of temperature, humidity, and other variables of the atmosphere. The result was the seminal "Nakaya Diagram," a graphic illustration of what types of crystals—from dendrites, to needles, to hexagonal plates—form in certain atmospheric conditions. As Bentley had hypothesized, Nakaya ascertained that temperature is the key to crystal formation. The dendrites, those six-sided flakes in the idealized form, develop when humidities—or dew points—are high and cloud temperatures are about 5 below zero Fahrenheit, he discovered. Temperatures in the 20s favor hexagonal columns that don't accumulate as efficiently. The lower the humidity, he found, the simpler the crystal. Conversely, higher humidity favored complexity and rapid growth. The harvest of his research was the world's first classification system for snowflakes, likened to a "Rosetta Stone" of snowflakes morphology.[45]

In the 1990s, in the early days of the internet, Libbrecht acquired a copy of Nakaya's *Snow Crystals: Natural and Artificial* when a publisher made a collection of rare books available. "That was just a great read," he said. When it was published in 1954, with 1,500 illustrations, Nakaya's book was another breakthrough in snow science. "No one can open the volume without some gain," *Science* magazine said in its review. Among its findings the book documented that the classic dendrite forms "constitute only a small part of the naturally formed types."[46]

Libbrecht said he had no idea that snow came in such tremendous varieties; that he had been missing out on a miniature world; and that a snow "flake" that resembled a descending bullet actually was a "capped column," a three-dimensional hexagonal solid "capped" at either end by stop sign–shaped plates. "It bothered me that I didn't know," Libbrecht said.[47] Physicists are like that, too.

For Nayaka, the research was as much about the intrinsic beauty of the snow, itself, as it was about the science. "Snow crystals," Nayaka wrote in 1939, "are letters from heaven."[48] That poetical relationship and mystical association are leitmotifs among the generations of snow's admirers. Perhaps Thoreau was speaking directly to the likes of Nakaya and Libbrecht when he said of snowflakes, "*I should hardly admire more if real stars fell and lodged on my coat.*"[49]

Cosmology and the Snowflake

If the species produced a less likely candidate than a Vermont homeschooled farmer to advance snowflake science, I would cast my vote for Ken Libbrecht.

I had the good fortune of becoming acquainted with him because we share an interest in snow, albeit in quite dissimilar ways. Had I known anything about astrophysics (remember, I warned you that knowing too much will never be my handicap) I would have been aware that he is one of the world's most renowned scientists. You would not guess this upon meeting him. He is unassuming; has a finely honed sense of humor and an engaging skepticism; and still has a trace in his voice of his native North Dakota, where he once had the thrill of frolicking in house-high snow, a la the Barry Burnell alleged experience in Michigan to which I refer in the introduction.[50]

Libbrecht's research focus is cosmology. Carrying on the universe-redefining work of Albert Einstein, he is an internationally renowned expert on gravitational waves and laser interferometers. For academic journals he has coauthored such page-turners as "Coulomb Distortion of Pion Spectra from Heavy-Ion Collisions" and "Facular Influences on the Apparent Solar Shape." That doesn't seem like the stuff of a snowflake connoisseur.

One would assume that such a snow-crystal epicure would be a weather enthusiast. At least, I assumed so. I was wrong, but then again in Pasadena, Libbrecht says, there isn't much weather to be enthused about. I was surprised when he told me that he had found the weather app on his phone to be reliable, his only complaint being that he was frustrated it didn't give him yesterday's highs and lows. Actually, I was surprised that a scientist of his sophistication would bother with a weather app. I don't find them particularly reliable about what's happening *now*, let alone in the future. I once asked a meteorologist who worked with those things for a living how they could be wrong about the current weather. He provided one of the more existential answers I've encountered in my career. He explained to me that I had to keep in mind that *now* is a forecast. As for yesterday's weather, one doesn't need a weather app or a weatherman to know what happened yesterday. Climate data is readily available in a variety of places, on the National Weather Service and commercial-service sites, and in something called a newspaper. I am aware of at least two incredible snowstorms that Ken Libbrecht experienced, but they don't seem to have made much of an imprint on his memory.

And, of course, we have that other incongruity: The polar vortex hasn't shown much interest in Pasadena, which has an annual snowfall about 200 inches less than Sapporo's. The sun shines 292 days a year in Pasadena, and its annual parade and subsequent college-football bowl game on January 1 have never been snowed out. The average high in the dead of winter is close to room temperature.

But Ken Libbrecht appears to have developed a highly sophisticated taste for the incongruous. In the scientific community he quite proudly is the ultimate flake, unashamedly mesmerized by what he calls "these miniature ice masterpieces." He reminds everyone that not a penny of taxpayer money has gone to this research.[51] Without apology, Libbrecht has devoted a decent share of his professional life since the mid-1990s to finding out as much as he can about them, although they have been chary with some of their basic secrets: Why do they assume certain forms and different temperatures, and, more fundamentally, why are they always six-sided, whether they are the classic stellar dendrites, or tiny "needles," or capped columns? He hasn't learned everything, but he is confident that in this field, at least, he is ahead of his peers. Whenever he talks to other scientists, he says, he relearns that they "know nothing" about snow crystals.[52]

His journey to the universe of the snow crystal has been appropriately nonlinear, like everything else that involves the atmosphere. He grew up in Fargo, the site of a 1966 blizzard captured in a legendary image of a man standing atop snow heaped to the top of a utility pole. At the time, Libbrecht was eight, and one would think that a snow memory at that age would resist all melting. Yet, when I mentioned that storm, it took a moment for the memory to condense, and then, yes, come to think of it, he did recall sliding down the roof of a shed onto the snow below, and yes, that had to be the storm. And that is what brought to mind Barry Burnell. His academic record suggests he always was a person of prodigious intelligence with a restless mind. He was accepted to the California Institute of Technology and quickly impressed his professors. He was a whiz in physics, graduated number three in his class, and was accepted into the doctoral program at Princeton. In that storied central New Jersey town, he was fortunate enough to experience a spectacular snow event.

Before he received his Ph.D., he witnessed the attack of the dendrites during the East Coast blizzard of February 11, 1983. Princeton and its stately buildings were silenced under two feet of masterfully wind-sculpted snow. The flakes obscured the air so blindingly and accumulated so rapidly on that Friday afternoon that motorists were stranded on the New Jersey Turnpike and stuck in the Lincoln Tunnel under the Hudson River for eight hours as snow fell at the rate of as many as five inches an hour. The "gravity wave" that ignited the blizzard's incredible snowfall rates along the I-95 corridor received a special mention in the seminal *Northeastern Snowstorms*, by winter-storm guru Louis W. Uccellini, now head of the National Weather Service, and colleague Paul Kocin.[53] What Libbrecht remembered weren't the gravity waves or much else about the storm, only the fact that his car was so buried in snow he decided to leave it where it was until spring.

If the blizzard rekindled treasured memories of North Dakota winters, it did not dissuade Libbrecht from returning to the Land that Winter Forgot to become an assistant professor at Caltech. He was promoted to associate professor in 1989. For his pioneering research on the sun, in 1991 he was awarded the Newton Lacy Pierce Prize in Astrophysics. Six years later, he became the executive officer of one of the world's most celebrated physics departments.[54]

He was an academic superstar, and he was about to gain fame in a fashion that no one in the Caltech physics department nor Libbrecht himself could have forecast. It all seemed so much like a cosmic joke that only someone comfortable with nonlinear chaotic systems could apprehend.

"Most physicists want to work on quantum field theory and theory of everything. Well, that's pretty hard to do!" Libbrecht has said. "First of all, some of the smartest people in the planet are working on that, so if you want to make a contribution, that's tough."[55]

He still is very much participating in "serious" research, the LIGO project, for example. "Cosmology is about the universe, and there's no question that is good stuff," he said. "At the same time, I don't think everyone should do that. I've worked in fields where I've felt that if I didn't get up in the morning and come in and do whatever I had to do, somebody else would just do it."

In the case of LIGO, he said, "I am a small player on a big project—it involves around 300 researchers worldwide." LIGO is classic "big science" that requires big money. Snowflake research by contrast is "almost an anti-LIGO." It doesn't require great funding resources, just Ken Libbrecht's Visa card.

He discovered another bonus when he began his investigation of snowflakes: He had the field virtually to himself. "I found there was hardly anybody doing anything," he commented.[56]

So just how and why did he get into it? He said that he and a postdoctoral student were discussing ways of doing something "new and interesting." How about looking at the physics of crystal growth, which led to the question of which crystals to grow. Silicon? No, that already was overdone. Then came that *eureka!* Why not try to grow ice from water vapor, and isn't that how snowflakes were born?[57]

Libbrecht has photographic skills, his protestations to the contrary. He told me that his real ambition in life was to be a graphic illustrator; Hooke showed that a scientist could be quite a good one. Libbrecht said he abandoned that path for himself on the ground that he couldn't draw. But, yes, he can handle a camera. To capture flakes, he has taken photographic excursions into the frostbite zones. One of his favorite locations is in northern Ontario. "All the good snowflakes are on the edge of civilization," he said. The exception would be Fairbanks, where some humans actually live and work in temperatures that Libbrecht could not fathom tolerating. If you really want an otherworldly experience, he said, "You don't have to go to another planet. Go to Fairbanks."[58]

He has acquired outstanding samples in the frozen wilds, as had Nakaya. But like Nakaya, to understand crystal growth, he needed control of the process. He needed to find a way to make his own crystals, to photograph them to analyze them. He was about to become the Nakaya–Bentley.

One thing about those laboratory snowmaking devices, they are not sold in stores, nor can you find a handy 3D manual for easy assembly. Libbrecht took the molecular approach, as in assembling his apparatus piece by piece, and he does happen to be a big fan of molecules because, as he says, that's what stuff is made of.

His first crystal-maker was a vapor diffusion chamber, constructed out of plexiglass and Styrofoam, and Ken Libbrecht would become an important consumer of Styrofoam. It was an opaque box about two feet high and one foot wide. The box was insulated to maintain cold on the bottom, with temperatures of about 40 below Fahrenheit, and the heat at the top, about 104 degrees, or what it might feel like during a Pasadena heat wave. After a small quantity of water was placed at the top, vapor diffused through the box and condensed in growing ice crystals as it came in contact with the cooler air. Libbrecht and an undergraduate made the first "designer" snowflakes in the summer of 1997. Libbrecht attached the lens of a high-powered microscope to an opening in the chamber. He projected super-enlarged, magnified images of the crystals onto a TV monitor.

Libbrecht and his associates became more efficient at the snowmaking business, and in 1999, when Libbrecht discovered a 1963 paper by a meteorologist, Basil Mason, the result was a production breakthrough. Following Mason's procedure, he inserted a wire into the bottom of the diffusion chamber and applied 2,000 volts of electricity. The electrical jolts generated blizzards of potential crystal masterpieces to catalog, to analyze, to photograph, to admire.

Throughout the years Libbrecht has constructed different types of snowmakers, among them a refrigerator-size "convection chamber," basically a cold oven that bakes snowflakes. His snow lab, across the corridor from his office, located in a muted and tasteful building in the heart of the Caltech campus, is a visual counterpoint to the monolithic tower next door housing the Milliken Library. The lab, in turn, would be a visual counterpoint to the building housing the physicists, an assemblage that evokes a high-tech version of hay and baling wire. It actually occupies two rooms. At least it did when I was there.

Two chambers dominated the main room, a rectangular one about the size of a hotel-room refrigerator and a circular one, roughly the size of a keg of beer. Both were encased in Styrofoam bound with FlexFix tape; the circular one could have passed for a giant vat of takeout coffee. Connected to the machines were video monitors, a printer, blue and black plastic tubing, and more wires and connectors than I've seen in hospital intensive-care wards. Taken together, the entire room could constitute a priceless work of modern art. I've seen far lesser works in art museums.

Those machines have served snow science and Libbrecht more than well. Libbrecht shared his first photographic images publicly in 1998. He recalled that at the time it seemed that everyone was creating websites, so why not him. He created one for his snow-crystal images. He was nonplussed by its popularity. Among the things with which he is unfamiliar are the workings of the media. Reporters are always looking for fresh ways to frame familiar stories, and having a site devoted to snowflake images and snow science was a godsend. I was among the reporters ecstatic to see such a resource and learn that it was maintained by an accessible scientist.

In 2003, his fame received a jolt of electricity, not unlike the one that the Basil Mason wire gave to his snowflakes. Collaborating with photographer Patricia Rasmussen he published *Snowflakes: Winter's Secret Beauty*. It was only 112 pages and mostly images. Libbrecht, like Bentley in his early years, protests that he is not a writer, but as in the case of Bentley, the evidence suggests otherwise.

A review in the *American Journal of Physics* called it a "magnificent little volume."[59] The book was such a hit that Libbrecht followed up with several sequels. With the book revenues he was able to buy more Styrofoam, tape, wires, and visual equipment. Three years after that first book was published, the U.S. Postal Service printed 3 billion first-class stamps with his snowflake images. They sold out.

The Swedish government gave him the Lennart Nilsson Award, the citation declaring that with his photographs, "He turns mathematics, physics, and chemistry into images of great beauty."

The Swedish postal service also used his images on stamps, choosing them among 50,000 applicants for that honor.

One would think that Caltech would hold that such visibility would be a boon to the institute's brand. Libbrecht says, well, no; a nice Swedish Nobel

Prize would be more welcomed by Caltech than the Swedish Lennart Nilsson Award. In due time, he confided, the institute would want his lab space.

He will still have his office, with a window onto the campus. Its shelves hold books with unpronounceable names interspersed with his snowflake books. Across from the shelves, one item stands out on his bulletin board, a print of a publicity photo from the Disney movie *Frozen*. The Disney people had come across his work and hired him as a consultant. They were determined to portray snowflakes as realistically as possible. Libbrecht worked with them and stressed that above all, the flakes had to be hexagonal. Indeed, the one in the poster emanating from the hand of Princess Elsa is quite a realistic-looking dendrite with six fern-like arms.

Libbrecht admitted he was thrilled when it won the Oscar for best animated feature, and when the movie came to his local theater, he was anxious to see it. He took the wife and their two offspring, and truth be told, he was more interested in the credits than the plot. The movie ended, and the credits started rolling. Everyone else began leaving the theater. The Libbrechts didn't budge. He said the credits kept rolling for what seemed like 20 minutes. The theater lights came on, everyone else had left, they were the only survivors. They were rewarded; the family hero's name scrolled up the screen.[60]

Disney didn't pay him much, just $200, but his snow research has been financially rewarding. When he gives talks, he says he has discovered one practical application of his work: In addition to purchasing equipment, his book revenues have helped pay for college for his children.

A Unique Challenge

Libbrecht has resolved one riddle of consequence regarding snow. Bentley's assertion about the uniqueness of the snowflake had been challenged before, and not just in the realm of common observation: Anyone who has inspected flakes during a snowfall could not be faulted for questioning Bentley. In 1988, the concept was subjected to what appeared to be a scientific challenge. That year the journal *Nature* published a letter from the late Nancy Knight, a scientist at the National Center for Atmospheric Research who said that in the upper atmosphere she had found two snow crystals that indeed appeared to be identical. Ironically, at the time she was involved in global warming research. Her brief letter created a national sensation. Libbrecht remembers it well. I was among the reporters who attempted to reach her for elaboration. Her husband, a colleague and ice-crystal expert, intercepted my call and insisted on speaking on her behalf. He said she was taken aback that the letter was received so seriously; it was not a peer-reviewed academic paper. His wife was astonished by the reaction. Indeed, he said, it might be all but impossible to find two identical snowflakes. Although he has made crystals in the lab quite close to identical, Libbrecht concurred.

How could that be? A snow crystal, explains Libbrecht, contains about a quintillion—that's 10 to the 18th power in scientific notation—water molecules. We will spare our readers the zeros. About a quadrillion of those—10 to the 15th power—are scattered randomly throughout the six-sided structure. It is theoretically possible that a particular molecular arrangement in one crystal could be replicated, although that is the longest of long shots. Said Libbrecht, "The odds of it happening within the lifetime of the Universe is indistinguishable from zero."[61]

He is talking about the individual snow crystal. The odds would approach zero asymptotically for the crystal aggregations that in general usage are called "snowflakes." Those saucer-size flakes that fall when temperatures are near freezing might actually be clusters of hundreds of crystals.

After landing, crystals and clusters rapidly lose what remains of their exquisite forms and, like so many voices blending in a choir, assume anonymous roles in contributing to that uniform whiteness, or what we perceive as whiteness.

Carefully examine individual snowflakes and you might be surprised to discover that they aren't "white" per se. They are crystals of a high aesthetic order, like shards of Waterford. They are translucent, as salt or sugar grains, and accumulated snow appears white for the same reason that salt in a shaker or sugar in a bowl does: All light is reflected from the grains. Nature, the grand magician, holds the mirror as snow transforms the landscape. As the sun prepares to retire for the day and its rays become more oblique, the crystalline qualities come out of hiding as the snow sparkles with the colors of the spectrum. That effect might well be what "Winter Wonderland" lyricist Richard B. Smith meant when he wrote, "In the lane, snow is glistening."

Snow's Wild Ride

Growing up in a gritty and grimy industrial town next to a rat-gray river, I have always been awed by freshly fallen snow's majestic powers to impose a sense of order and tranquility on any environment. Yet, one of the secrets unlocked by generations of observers, from the Scientific Age to the 21st century, is that a snowflake's career is the antithesis of tranquility and order.

Some snow crystals can trace their roots to the most cataclysmic explosions on the earth, volcanic eruptions, and others to the remnants of meteors in the form of disintegrated interstellar dust. The snowflake-building process begins when water vapor in a cloud attaches to microscopic particles, so-called condensation nuclei, in the high atmosphere and freezes, forming seed crystals. (Efforts to replicate that process through "cloud seeding" have yet to yield reliable results.) Building on Nakaya's work, Japanese colleague Motoi Kumai estimated that a single crystal could attach itself to 10 million such microscopic particles.[62] In addition to meteoric detritus and volcanic fallout, those particles could consist of the likes of airborne sea salt or pollen. As water vapor gloms onto the initial crystal, the added mass adds weight. When the growing crystal becomes heavy enough, it begins its long and stormy relationship with wind and gravity, and then its life becomes more complex and interesting. As Descartes astutely postulated, a crystal would be so light that it would be at the mercy of any wind current. Winds move horizontally, and vertically, and thus a crystal would encounter a tremendous variety of temperature and humidity conditions before it could begin that ponderous trip to Earth's surface. Those conditions would redefine the crystal as it encounters water-vapor droplets. Each attaches itself to the crystal, which continues to grow. As Libbrecht observes, water molecules naturally align themselves in a hexagonal pattern, and as they attach to the crystal, the pattern is repeated symmetrically, extending the six "arms" radiating from the nucleus.

When the seed crystal accumulates enough allies to become sufficiently enlarged, it can be called a "snow crystal," or the "snowflake" in its proper form. It is estimated that the earthward journey can take as many as three hours, depending on multiple factors, which include winds, flake mass, and "drag" (which considers the density of the air). A flake can get blown sideways a tremendous distance, and that's one reason that on occasion snow can be falling even as the sky directly above is bright blue. Untold quadrillions will never make it to Earth's surface alive, evaporating or sublimating and assuming a vaporous state as "virga." When virga is present in the atmosphere over a given area, radar images can give the appearance that it is snowing. That is an illusion, however; the snow is not reaching the ground.

Flakes that remain intact attract peers and are no longer individual crystals but aggregations. Especially during late-winter storms, when temperatures are close to freezing and dewpoints are higher than they would be when it was colder, those aggregated flakes can take on outsize proportions, rivaling espresso saucers. Since they are so weighty, they gain speed and literally constitute "heavy snow."

As Bentley verified, under a microscope a snowflake bares the evidence of what it has been through on its raucous trip through the atmosphere. The seed crystal began as a simple six-sided plate. The arms grew as the crystal migrated through the clouds, responding to changes in temperatures and dew points. The arms maintained symmetry because even though they grew independently, they all experienced those same changing conditions.

Each crystal follows a unique path, so each undergoes a different set of conditions. That path never will be replicated precisely. Nor will the snow crystal, something that Bentley concluded after examining all the available visual evidence. "Every crystal was a masterpiece of design," he declared, "and no one design was ever repeated."[63]

Bentley's documentary evidence and insights were groundbreaking. They merited far more comment and posthumous acceptance in the scientific community than they elicited in Bentley's lifetime, in Blanchard's view. The biographer, holder of a doctorate from MIT, wondered if a certain credential snobbery was at work, with Ph.D.s disdaining the labors of a mere farmer. Blanchard said it also was possible that in the academic community Bentley's enthusiastic prose violated certain professional taboos. How many academic papers contain that phrase "cloudland"?

In his later years, Bentley did undertake public lectures at scientific institutions, speaking to groups at the Buffalo Museum of Science and Philadelphia's Franklin Institute. He also lectured in his hometown of Jericho, where crowd control was never an issue in a village where farmers, like Bentley's own father, wondered how a healthy man could waste time looking at something as pointless as snowflakes.

"Oh, I guess they've always believed I was crazy, or a fool, or both," Bentley said.

Years ago, I thought they might feel different if they understood what I was doing. I thought they might be *glad* to understand. So I announced that I would give a talk in the village and show lantern slides of my pictures. They are beautiful, you know, marvelously beautiful on the screen. But when the night came for my lecture just six people were there to hear me. It was free, mind you! And it was a fine, pleasant evening, too. But they weren't interested.[64]

"He wanted everyone to see what he saw," said Richardson. "It wasn't about money for him." He did sell some of his photographs—for a nickel each—to cover his costs, which he never came close to doing.[65]

Bentley accepted the townsfolks' lack of interest in his "marvelously beautiful" treasures unresentfully.

"They thought he was a little cracked. Those were his words," Sue Richardson said. Some of his brother's eight children, with whom he shared the house, came to appreciate his accomplishments and their peculiar uncle.

Among them was Sue Richardson's grandmother, Bentley's favorite niece.[66] Bentley could not have known that one day his handiwork would be the village's prime tourist attraction, the Snowflake Museum. It is housed in an unassuming and quintessentially Vermont red farmhouse.

Unlike colonial predecessors Increase and Cotton Mather, Bentley saw the benign hand of God in winter's fury. "The snow crystals come to us not only to reveal the wondrous beauty of the minute in Nature, but to teach us that all earthly beauty is transient and must soon fade away." That is the concept captured eloquently and succinctly in Robert Frost's "Nothing Gold Can Stay," one of my favorite poems, but Bentley continued with a note of supreme optimism. "But though the beauty of snow is evanescent, like the beauties of the autumn or the evening sky, it fades but to come again."[67]

The return of that "beauty" hasn't been universally embraced in the 21st century, but snow wasn't always cast as the nemesis of the business of life in the United States. In the decades between the American and Industrial revolutions, snow assumed another role: It often was a welcome visitor to the emerging nation that it helped to build.

CHAPTER THREE

Pax Nivalis

When Snow Was Welcome

In Bentley's lifetime, America's relationship with snow became evermore adversarial, culminating in an event that forever changed the perceptions of snow in the United States. Its tremendous recreational and aesthetical aspects notwithstanding, in the 21st century snow is viewed as the great enemy of commerce, if not public safety. Throughout the years, the government winter-storm warning language has become only more foreboding, the lock-up-the-children advice more ominous.

Thus, I was surprised to find out that in the period between the American and Industrial revolutions, the relationship between snow and the transplanted Europeans evidently had experienced a decades-long thaw. Like a snow cover in April, the colonial-era writings so rich with evocations of Divine wrath and the "Hands of Providence" vanish as the colonial era yielded to the nascent republic.

Perhaps the disappearance of the Providential references was an Enlightenment-era byproduct. Or perhaps the spirit of accommodation was the result of two or three generations of wintry experience. Or welled from observing the winter-savvy North American indigenous peoples, whose ancestors had weathered the local winters for thousands of years. One well-accepted hypothesis holds that those ancestors had migrated from Northern Asia, crossing what is now the Bering Sea, when its waters were locked in ice sheets, the settlements progressing southward. If that were the case, those people would have had a lengthy history of confronting the worst that nature had to offer. The Europeans likely learned from them invaluable survival

35

strategies, information about how to protect themselves from the harshest conditions. The McCord Museum in Montreal, where winter winds bite and the darkened, solidified ice patches resemble the backs of frozen whales, has wonderful examples of footwear and outer garments, crafted from Caribou fur, worn by Native Americans.

The early settlers did demonstrate an impressive willingness and ability to adapt to challenging conditions, and to learn and prosper from the challenges, a skill that Frederick Jackson Turner extolled in his *Frontier Thesis*.

For whatever reasons that might explain the phenomenon: More than a mere acceptance, the snowy realities of living to the west of the Atlantic were embraced by the new Americans living in snow country, and exploited, even enjoyed. In the words of historian Blake McKelvey, "Towns and cities acquired a merry atmosphere in winter seasons."[1]

One obvious factor in what I call the "Pax Nivalis" was the fact that the national economy was in its barest infancy. With the British occupation, during the Revolutionary War New York City's population had dropped to less than 15,000. After the war finally ended the newly created 13 states generally were settled sparsely.

Snow was a boon to the new nation's growth.

Right after the Revolution and well into the early 19th century, snow was commonly viewed as the valuable ally of commerce, contributing to the development of inland towns. The cities in the coastal plain—Washington, Philadelphia, New York, Boston—experienced their shares of moderate and mega-storms; however, accumulations away from the coast, where the land begins its gradual upsloping, would tend to be significantly higher. That would be particularly true along and above the elevated "fall line," which parallels what is today the I-95 corridor. This phenomenon is referenced in those winter forecasts for the East Coast cities that call for two to four inches in the city and four to six in the northern and western suburbs. Temperatures decrease with elevation, which also can exert a "lifting" effect to rising air currents that condense into precipitation, thus enhancing snowfall. Plus, in those areas the snow and ice cover would have been more extensive and longer-lasting than it would have been in those urbanizing areas of the coastal plain. Snow has a refrigerating effect and is an excellent sun repellent.

"New communications are open over lakes and rivers through forests hitherto impassable," Hector St. John de Crevecoeur wrote in his *Letters from an American Farmer*, a series of letters that provided Europeans with some of their first insights into life in the New World. They were not published until 1792, but the transplanted Frenchman who worked a farm in upstate New York had written them several years before the Revolution. He is not as well-known as fellow Frenchman de Tocqueville; however, de Crevecoeur's writings were justifiably heralded by his Old World contemporaries and were immensely popular.

His observations about snow and winter might be unrecognizable to the modern reader. "Immense is the value of this season," he said. "Logs for future building are easily drawn to the saw mills. Ready-piled stones are with equal ease brought to the intended spot." Sleds and sleighs assumed the roles of modern-day 18-wheelers, carrying cargo across snow-and-ice surfaces and reliably frozen waterways.[2]

In the snowier regions, rather than remove it, the locals used massive, horse- or ox-drawn wooden snow rollers, weighted with ballast, to smooth it. Some samples of these almost cartoonish objects, which resemble a cross between old-fashioned water wheels and oversize modern-day steamrollers, survive today, although they are rarities. The village of Jackson, New Hampshire, about 10 miles south of Mount Washington, has two sets of rollers on display in the center of town.

Contrary to today's concept of winter street maintenance, early 19th-century road crews had an inverse approach, treating the surfaces as modern resort operators would treat popular ski trails, grooming them and spreading snow over bare spots, like lathering frosting over bare spots on a cake, to maintain friction-free continuity.

Rather than inducing a craving for isolation, such a tamed wintry landscape constituted a visual banquet that would stimulate the appetite to socialize. A generous well-packed snow cover was a welcome source of diversion and recreation, observed McKelvey in his *Snow in the Cities*.

Rather than inhibit travel, it promoted it.

> The delights of gliding over the snow-packed streets and rural roads prompted residents of towns and cities to defer distant journeys until the snows arrived. . . . After each storm subsided, many joined their neighbors in hitching horses to heavy sleds to break open the roads for merry and jingly sleigh rides about town.

With snow came opportunity, not hardship. Said de Crevecoeur: "The convenience of traveling invites the whole country to society, pleasure, and visiting."

By all available accounts, the citizens of the new nation had reached an accord with nature during the first three decades of the 1800s. After a January 1830 snowfall, the *Boston Gazette* wrote glowingly of the Boston variant of the snow-grooming practice.

Several sleds filled with men and boys, attached to each of which were six horses, were driven through the streets of the city yesterday afternoon to break and level the drifts. Our truckmen understand this business, they practice the leveling system, on all great occasions, with more success than any other class of citizens.[3]

The end of the Great Thaw didn't occur precipitously or uniformly in the regions affected by winter storms, but the year 1830 is of particular significance. With increasing population densities and economic ambitions, tensions between snow and the emerging nation were inevitable. In the end, this Pax Nivalis was as doomed as a snowpack subjected to sublimating winds and the increasing solar power of spring. Such was the price of progress that attitudes toward snow would undergo an evolution that is still unfolding today. As so many changes of consequence, as the atmosphere, itself, this one was by no means linear. The catalyst was another cosmic, nonlinear development, the Industrial Revolution.

Railing against Nature

Snow so pure when it falls from the sky,
To be trampled in mud by the crowd rushing by;
To be trampled and tracked by the thousands of feet
Till it blends with the horrible filth in the street.—James W. Watson[4]

The coming storms of tensions between winter and American urbanites incubated in the 1820s. At the time, Baltimore was the third-largest city in the country, with a population rivaling that of number-two Philadelphia, 90 miles to the north. With its excellent and well-sheltered harbor Baltimore was undergoing a growth spurt. So were its financial fortunes, and in that respect it was not unique. Competition from the north was intensifying, from not only Philadelphia, but also an emerging economic behemoth farther to the north.

One of the great engineering projects in history, the digging of the Erie Canal, was underway in New York state. It would dramatically change the fortunes of New York City, which was making a rapid recovery from its Revolutionary-era torpor. Less than 50 years after it was abandoned by the British, it would become the first U.S. city with a population of 200,000. It was hardly a portrait of city planning at the turn of the century. The population was overwhelmingly concentrated in Lower Manhattan, where the layout of the serpentine streets resembled that of a medieval town. Appropriately or coincidentally, it was in 1807 that Washington Irving satirically branded the city "Gotham," or "Goat's Town," a name drawn from folktales of the Middle Ages.[5] New York's city council had more-modern ideas, although the moniker "Gotham" would stick. My guess is that most people who rather proudly use that term are unaware of its origins, as were the creators of DC Comics, who assigned Batman to Gotham City.

Coincidentally it was in 1807 that, at the city council's behest, Gouverneur Morris and two surveyors became the "commissioners of streets and roads." They spent four years developing an ambitious vision for the rapidly growing metropolis, which grew to become the nation's biggest city in 1810. To encourage the population to inhabit the more northerly reaches of the 13.4-mile-long island, the three presented a radical plan in 1811, to create wider, north-reaching avenues, crossed by streets that extended river to river, creating a classic grid pattern.[6] Lower Manhattan wasn't going to be able to hold everyone who was showing up. Given its location, the city was a logical port of entry for continuous influxes of immigrants and goods.

And that was before the Erie Canal.

It opened in October 1828, connecting Albany with Buffalo and, more significantly, the entire Hudson Valley with the Great Lakes markets. The canal would assume an indirect but important role in the future of snow removal. With its opening, New York City became the southeastern terminus of a waterway coursing through towns, settlements, and sources of valuable markets and raw material in the new nation's interior. As Robin Nagle notes in *Picking Up*, via the canal a ton of flour could be shipped from Buffalo to New York City at one-seventh the cost of transporting it over land. Towns along the canal route, which includes the New York Snow Belt, grew. New York City didn't; it exploded.[7]

To the south, the city on the Chesapeake Bay was preparing an innovative engineering strategy of its own to exploit the inland potential. Two months before the canal opened, Charles Carroll, the last surviving signer of the Declaration of Independence, with a shovelful of dirt, symbolically began the construction of the nation's first commercial railway line, the Baltimore and Ohio. Although its ultimate ambition was to mine those "western" markets, the first section spanned just 13 miles, terminating in Ellicott Mills, Maryland. The rail opened for business on May 24, its first runs being horse-drawn, with a one-way trip taking about 90 minutes, traveling less than nine miles per hour; however, the nation's first passenger rail service was a sensation, and the company had to hustle to get more cars built to keep up with demand. The rail grossed almost $15,000 in its first year of operation.[8]

That initial portion of the rail, completed in May 1830, was destined to become a Bunker Hill in the history of snow-fighting in the United States, and the B&O the venue for one of the first wintry shots fired against nature in the Industrial Revolution. In the summer of 1830, inventor Peter Cooper, who had made a fortune as a glue manufacturer in New York, developed a coal-powered prototype engine, the "Tom Thumb," with scrap iron and "musket barrels for boiling tubes." With B&O directors aboard, it was able to make the 13-mile run in 57 minutes. It later famously lost a race to a horse-drawn car, but the directors were "convinced" that "they should turn to steam."[9]

On the weekend of January 14–15, 1831, a prodigious storm with ferocious winds lashed the entire east from the Gulf of Mexico to Maine. For intensity and geographic reach, it surpassed any winter event that had occurred since Europeans had arrived on the continent, according to David Ludlum.[10] It layered 30 inches of civilization-stopping snow atop Gettysburg and 22 on Pittsburgh. Drifts as high as 12 feet shut off mail service and halted stagecoaches at Lancaster. The nor'easter was the "greatest snowstorm of the 19th century along the Atlantic seaboard."[11] For areal coverage, it surpassed even the March 1993 Superstorm.[12] In Baltimore, this was the deepest snow in more than 70 years. The brand-new railway, the subject of widespread skepticism about whether this radically new form of transportation could navigate through the worst of winter, was stopped in its tracks.

The setback was temporary. The B&O responded ingeniously. To push aside the snow, it deployed a device invented by a "Mr. Winans," according to the *American Journal of Science*. It would become a template for generations of variants. The contraption was shaped like a three-dimensional directional arrow and pulled by several horses "at a trot." It had a wedge of iron at the front that acted more or less like a snowplow, although that term would not be invoked for more than a decade. The wedge was trailed by metal runners, parallel to and above the rails. Between the rails and the runners were angled iron teeth for scraping off the ice and snow. This particular plow-like variant was a first of its kind, and it is uncertain whether it was ever replicated. It cleared the rail sufficiently to allow passage for locomotives.[13] For the B&O this novelty drew a welcomed wave of publicity to a burgeoning enterprise that would re-make the nation.

The railroad industry was at the vanguard of the country's changing relationship with snow. More than with any other development, with the rails came the era of snow-fighting in the United States. Very simply, snow was their enemy, and intercity rail systems were proliferating as the 19th-century population virus spread from the East to the Ohio Valley and beyond, in places where wintry conditions could occur during a third to a half of the year. A vast and intricate intracity rail network was developing in New York, where the population had topped 500,000 by 1850. With the rails came new efficiencies and new conflicts in the city that would become the nexus of a seminal event in the history of snow-fighting.

When Snow Becomes "Refuse"

All the population centers in the Mid-Atlantic, Northeast, and broadly defined West had to contend with snow. While it remained a source of enchantment in the countryside, in the cities it was assuming a new career as "refuse." So observed Bernard Mergen, who was an American studies specialist at George Washington University. Urban snow tensions flared in the likes of St. Paul, Minnesota, and Rochester, New York, and to the nth power in New York City. What made the snow-as-refuse phenomenon so problematical in Gotham was the almost otherworldly pace of population and commercial growth. Washington Irving wouldn't have recognized the place.

With the expansion came otherworldly quantities of garbage and assorted refuse, which included stinking animal carcasses. Some streets in Lower Manhattan were becoming cesspools, notorious for odors that rivaled the local politics that allowed the accumulation of the trash and food detritus. In the words of a *New York Times* article, "Portions of animal remains, straw, excrements, filth of almost every description . . . cause a pestilential stench."[14]

The unsanitary conditions likely contributed to the fact that New York had one of the world's highest mortality rates among the world's biggest cities. One study determined that basic sanitation practices could have cut the 1851 death rate in New York by a third.[15] Snow might provide a temporary and transformative entombment of the garbage and decaying animal corpses rotting in the streets; however, the days when a snow cover could peacefully conquer that rancid accumulation were ending.

The use of steel tracks was becoming popular throughout the country, even in the absence of the iron horses. In New York, as the population swelled, rails appeared evermore frequently on the major Manhattan boulevards to accommodate horse-drawn cars. Pulling wheeled carts on metal rails, the horses were able to haul heavier loads and move more quickly than they could on conventional cobblestone or dirt streets. Eventually, steam-powered locomotives would replace the horses. The rails were boons to commuters, commerce, and especially the transportation companies, which had no time to stop for snow.

The reports of snow blockades on "steam roads" mounted. Those block-ades affected the New Haven & Hartford rail in 1839, and in 1843, when the Auburn & Syracuse was smothered by 25-foot drifts. What was described as an "ocean of snow" extended as far south as Philadelphia.[16]

Send in the Plows

The sea-changing weapon in the annals of Homo sapiens versus winter was the plow. It inevitably became the crucial link in the chain of events that would lead to government-operated and organized snow removal. It is un-clear whether the B&O's snow-mover was the model, but the rail companies unquestionably were the prime drivers of the technology and the prolifera-tion of plow use throughout the country.

As with the microscope, it is impossible to identify the person or persons who invented the plow or, for that matter, which device first earned the "snowplow" label. At least one patent application was registered in 1846 in England,[17] and several more patent requests were filed a decade later.[18] The date of the first documented use of a plow on a city street likewise is uncer-tain. A *New York Times* article after a brutal storm in January 1856 notes that the rail companies' only option was to shovel snow to the sides of their tracks. The following winter, however, after a powerful and frigid snowstorm had punished the dense and growing corridor from Washington to Boston with one to two feet of snow, a January 20, 1857, *Times* piece mentioned that plows were commissioned to move the white masses from the horsecar rail tracks on Sixth and Eighth Avenues.[19]

An article in March following a late-season heavy snowfall gave more detailed accounts of plowing along several avenues, including Sixth and Eighth. "During the day, a snow-plough was in operation on both roads," it said. On Second Avenue, "Every exertion was made to keep the tracks in or-der by using a snowplow and"—foreshadowing a popular modern strategy— "sprinkling salt on it." Salt also was spread on Third Avenue to promote melting.[20] Salt offered two important advantages in the developing war on urban snow and ice. The use of salt has become a popular winter weapon, for good reason. It is plentiful, the remnants of dead seas have left a virtually in-exhaustible supply, and it works. It quite effectively lowers the temperatures at which water freezes.

The melting powers of the salt that was spread on such thoroughfares as Broadway, which at times truly would become the Great White Way, ben-efited the wheeled vehicles. Conversely, it was the mortal enemy of those designed to glide on blades. One frustrated local publicly expressed indignation in a letter to the *Times* on January 5, 1859, complaining that the salt was ruining the street surface for sleigh traffic. No one could accuse the editor of pandering to the reader. The editor responded that the wheels of progress had a far higher priority than the metal blades of the city's past. The *Times* position notwithstanding, the use of salt remained controversial. On January 31, 1860, the paper carried an item noting that Common Council had appointed a committee to consider outlawing the practice.[21]

The war on snow was an imperative for commerce, and according to a piece in that January 5, 1859, issue of the *Times*, it was an ally in the war on poverty, providing economic dividends for the less fortunate. Snow meant opportunities to "thousands of poor people who seek employment and are glad to find it and to earn their half-dollar clearing away snow from sidewalks." The article continued, "It is not an exaggeration to say that yesterday, in New York and Brooklyn, some $30,000 were expended in this charity, and even a snowstorm, pitiless and merciless though it be, still brings a blessing to many a poverty-stricken outcast."[22] As in other cities, in New York shovelers heaped snow into horse-drawn carts that hauled away the piles. New York deposited its masses of snow into the East and Hudson Rivers.

With the intensifying efforts to unclog snow-choked arteries came collateral damage and complaints from some of the very people who stood to benefit. Those protests were attendant to the fallout from the fruits of the mighty plow, at once indispensable and an urban nemesis.

Unsurprisingly, the engines driving plow innovations were the railroad companies that were busily remaking the nation and its demographics with their steam-powered locomotives. In the arena of snow-moving, New York was more or less a caboose. Given the level of capital investment, the rails had to operate in all seasons through every kind of weather. If snow-fighting was all about getting the forces of obstruction off the tracks, the weapon of first and last resort was the plow. Rail lines cleared tracks with their massive, metal, wedge-shaped plows mounted to the fronts of engines. In a first-person account, a *Milwaukee Sentinel* reporter riding on a train so equipped described the experience as not unlike traveling on a white sea. From his vantage point, the clouds of snow raced past the front of the train "like foam from its bows."[23] These operations, however, weren't always smooth sailing.

Depending on the weight and depth of the snow, a rail company might have to use multiple locomotives, sometimes as many as 14, to propel a plow moving through the heaviest drifts.

The New York Central, on its upstate lines, opted for a sophisticated mechanism, invented in Philadelphia in 1857.[24] In addition to a snow-moving wedge, it was equipped with a broad scoop to deposit snow onto a conveyor that would load it onto a dump car. A decade later, a Canadian dentist, Jull Orange, would invent the rotary plow, the ancestor of the modern snowblower.[25] It had circular rotating blades that ingested the snow and then propelled it through a chute to either side of the train at the command of the operator.

The Urban Dilemma

The snow-moving wizardry developed by the railroads was an impracticality in the urban commercial corridors, where by the end of the 1860s the war on snow was exhibiting a certain ugliness to rival the hideous graying and blackening piles. The skirmishes coincided with complaints about weather-related delays in mail deliveries and produce shipments, and the general pace of the evermore bustling business of America. The urban plows that were so essential to snow-fighting arsenals were the sources of fresh and growing mounds of problems. They were deployed on ad hoc bases, used at the discretion of private interests. They were poorly regulated, if at all.

The melting was at least as disruptive as the storm itself, according to a *New York Times* account of a poststorm thaw. The slush was a "horrible black mass of mud and filth . . . sufficient to appall the stoutest pedestrian." Along the Eighth Street railway, "mounds and ridges of dirty" snow were "banked up." And on the Bleecker Street line, a "great sweeping machine, drawn by 10 horses, tore along, besmattering everybody as it passed."[26]

In New York, relations between the street-rail companies and the people who lived and ran businesses along their lines were on collision courses. An early indication of the simmering tensions came with the arrest in February 1867, of an employee of the Dry Dock and East Broadway Railroad Co., a Mr. John Loney, for plowing snow from Avenue B onto the sidewalk. Two winters later, the citizens along 34th Street between Broadway and Seventh Avenue extracted their own form of justice. After the 42nd Street Railroad Co. piled snow on their sidewalks, they deposited it right back on the tracks.

The growing frustrations of the populace were encapsulated in an 1871 letter to the New York City Board of Health in which the writer decried the "intolerable nuisance" created by a rail company's "using an immense snowplow to clear their tracks, by means of which slush and all kinds of filth was thrown over pedestrians and houses."[27]

A modest snowfall in December 1871, less than four inches, underscored the growing urban intolerance for nature's interruptions. The New York horse-car companies tried frantically to accommodate their growing ridership. But Henry Bergh, founder of the Society for the Prevention of Cruelty to Animals, blocked several of the overcrowded cars and demanded that the companies provide double teams to pull them. The particularly severe winter of 1872–1873 was replete with incidents of slips and falls, and broken bones, among humans and horses, the latter suffering execution for their pains.[28]

Judicial Intervention

That the plow conflicts would end up in court confrontations was inevitable. In a case that would have ramifications for every rail company and indeed the entire state, in 1875 New York City resident John Taylor Johnson asked the Common Pleas Court to issue an injunction to stop the Christopher and 10th Street Railroad Co. from clearing its tracks with plows and street sweepers. He contended that the obstructions caused by the rail's practices were preventing him from obeying a city ordinance to clear his own sidewalk. The case had added import, because at the time, the state legislature was considering a bill limiting the use of plows and sweepers.

The rail company argued that its charters compelled it to keep the trains running. Thus, it was required to clear the rails as quickly and efficiently as possible. No doubt to the relief of rail interests throughout the state, the judge ruled in favor of the 10th Street company. In turn, the company filed for an injunction of its own.

It asked a judge to order a cessation to the practice of throwing snow back onto its tracks. The judge reasoned that given that a court had ruled that the companies must honor their charters, it followed that they would have the "right to have the road left free." The rails had won another important round.[29]

Not all the litigation was as consequential. A gentleman named John Gray sued the Second Avenue Rail Co. for $500, claiming that one of its plows had injured him and his horse. Noting his careless behavior, however, the jury awarded him a grand total of six cents.

For the welfare of a city whose population would zoom past 1.2 million by 1880, the winter guerrilla activity had to cease. One important development was the decision by some rail companies to get above it all by building elevated railbeds for their steam-powered trains. A decade after the first experimental version of an elevated line operated on Greenwich Street from 1868 to 1870,[30] trains on raised railbeds were running along major thoroughfares, including Second, Third, Sixth, and Ninth Avenues, in the late 1870s. Obviously, that obviated the necessity for the rails to plow at street level, and the elevation would alleviate the problem of track-blocking drifts, but it would not resolve all the issues, as the rails would find out.

Back at ground level, the rising of the rails hardly eliminated the problems attendant to snow removal. Sooner or later, and assuredly it should have been sooner, New York City would have to impose order on snow-clearing operations. A crucial step in that direction was taken in 1881. Acting at the city's request, the state legislature voted to allow it to create a Department of Street Cleaning.[31] Among the duties it undertook was to oversee snow removal in the densest areas of Manhattan, from 14th Street south, organized into two sections, with Broadway as the dividing line. Under the regulations written by the commissioner, all plowing was subject to his approval. The rules expressly forbade the use of salt or any other melting material. They prohibited the plowing of snow onto sidewalks and curbstones. Snow was to be removed from all narrow streets.

While the city would now regulate snow removal, the plowing operations still were the province of commercial interests, not government, and the commissioner's edict applied only to Lower Manhattan.

An extreme act of nature, however, was about to alter snow-fighting forever. A new term had entered the American lexicon: "blizzard." In 1887, a blizzard had a cosmic and ultimately enduring impact on the sparsely settled West, a prequel to the one in the East that would redefine America's relationship with snow.

CHAPTER FOUR

Majestic Silence

The Storm That Changed America

No paths, no streets, no sidewalks, no light, no roads, no guests, no calls, no teams, no hacks, no trains, no moon, no meat, no milk, no paper, no mails, no news, no thing—but snow.—*Bellows Falls Gazette*[1]

On a Monday morning in mid-March 1888, not quite dawn, a 13-year-old girl who lives on the Lower East Side of Manhattan is standing by a window wondering if she has left Planet Earth. Meta Stern had gone to bed at the end of a rainy, dismal Sunday, resigned to confront another week of school in that time of year when the classes and teachers have long since passed the borders of tiresome, and it was too soon to dream of summer vacation. At least the *New York Herald* promised that the day would be sunny, and spring was less than 10 days away. Yet, on this Monday morning, her bedroom is suffused with an "unearthly brightness." It is not rain that she hears, rather "a heavy silence, . . . a deep, unbroken silence."

She perceives a subtle swishing, like the vague humming of a distant machine. She is drawn to the window, where the night before a "mournful" curtain of dismal drizzle veiled the view of Stuyvesant Park.

She opens the shade and gasps in disbelief. "I stood, motionless, speechless," she would later write, "for the sight I beheld was so different from anything my eyes had ever seen that it seemed to belong to another world."

The park is entombed by all-conquering mounds of snow. The iron gate in front of her house has lost half its height, and the fences that separate her yard from those of her friends have vanished. The trees are gossamer, ghostly figures swaying in a haunting gale. The snow appears to be rising out of the very ground as rapidly as it is cascading from the skies and blowing chaotically sideways. Earth and sky are conjoined.

She runs to her parents' bedroom and wakes Herbert and Gloria Stern, and tells them of a phenomenon she has encountered only in books that describe wild behavior of the untamed American West.

"Mother! Father! Wake up! A blizzard is raging!" Calling this a snowstorm, in her words, "would have seemed as absurd as calling the ocean a swimming pool." What she was witnessing was the Great White Hurricane of March 11–14, 1888, when house-high drifts smothered the nation's commercial capital.

Reflexively, Dr. Stern goes to his office to prepare for his patients, while his 13-year-old daughter wondered how she would get to school. But this is the morning of no paths, no streets, no trains, and assuredly no school. On a practical and more-urgent level, Gloria Stern grasps that this would be the morning of no meat, no milk, no paper, no mails. The Great White Hurricane has erased the buildings that were across the street on Sunday.

Nature has humbled the mightiest city in the Industrial World, isolating it from the rest of civilization, executing an all-out crippling assault at the dawn of the Machine Age. "The most amazing thing to the residents of this great city," the *New York Times* wrote, "must be the ease with which the elements were to overcome the boasted triumph of civilization." Or, as Meta Stern states simply, "The streets, the park, the city, belonged to the blizzard."[2]

For her, this was a magical victory, at least at first. For 400 others, and thousands of farm and feral animals, it was nature's fatal conquest. For some of Dr. Stern's patients, it was life-threatening, and for Dr. Stern's family the carnival of the snow would mutate into a high drama of maximum anguish.

Once he accepted that his patients wouldn't be able to come to his office, Stern made a commitment to go to them. He donned his "Arctic" gear and prepared to confront the frigid, swirling white masses. The front door was cemented shut by snow, but he was able to open the back door just enough so that his other daughter, Margaret, could help him create a path to the street. As he disappeared into the veiling curtains of snow, Meta and Gloria Stern feared he had just shoveled his way into his own grave.

The Millennial Storm

What the Sterns and 2 million Americans endured on March 11–14 was a historic siege of nature that for various reasons might qualify as the most significant and disruptive snowstorm in the nation's history.

That is a statement not easily elicited from a native of the Philadelphia city-state, where the concept of a New York–centric universe is usually an unbearably cold slap in the face. Yet, the preeminence of the Blizzard of 1888 withstands any regional prejudice. In terms of impacts, it was certifiably unprecedented, the Storm of the Century, the Storm of the Millennium, the Storm of Millennia.

That blizzards had raged across the continent for thousands of years is a certainty, although their occurrences weren't documented by the New World natives. Storms blow up along frontal boundaries. Even when ice sheets covered a majority of the North American land mass, those restless boundaries still would have existed. They likely would have been depressed southward, relative to where they are today, and crept slowly northward as ice melted. By the time European settlers arrived, the broad area that would become the northern two-thirds of the contiguous 48 states would have been situated under those frequent frontal battlegrounds and caught in the spectacular cross fire.

Frigid upper-level winds from the northwest and vapor-rich air mined from the Atlantic were ideal blizzard conspirators along the Eastern seaboard.

The formation of the Great Lakes, a spectacular Ice Age aftermath, created a new generation of blizzard-makers as those free-flowing waters forced icy air parcels to rise and condense into snowfall resembling madly swirling flakes of coconut in a whirring Cuisinart. Europeans had witnessed nothing like these lake-effect blizzards.

In the Ohio and Mississippi valleys, jet-stream winds and fronts that were express rails for Gulf moisture were the agents provocateurs of snowstorms.

In the sparsely settled West, what are now called "Alberta Clippers," barreling from the Pacific and crossing the Canadian border, and storms rolling out of the Rockies like rock slides set off snows that were piled into monstrous drifts by marrow-chilling hurricane-force prairie winds. With nothing to obstruct their polar-air importing force, they plunged temperatures to depths unfamiliar to Easterners.

One such storm, on January 9, 1887, was so destructive to livestock it became known as the "Great Dieup," a morbid play on the term the "Great Roundup." According to accounts of that storm, more than 1 million cattle, 350,000 in Montana alone, perished. So widespread were the deaths that when spring came, drinking water was scarce because rotting carcasses were clogging rivers and streams. That storm forever changed ranching practices in the West.

The following winter was a particularly deadly one for the United States. On January 12, one of those horrifying prairie storms that appeared to erupt from nowhere killed hundreds of people and became a tragic chapter in the history of Nebraska. It was known as the "Schoolchildren's Blizzard" because some of the victims were pupils who had gone to school on a balmy morning only to be stranded in a storm that drove wind chills to 50 below zero.[3] Growing cities in the Midwest, including Milwaukee and St. Paul, were crowned with as much as two feet of snow.

The actual death toll from those wintry assaults remains unknown, but it likely was higher than the more-famous storm that followed two months later.

About "Blizzards"

In all likelihood those storms in 1887 and January 1888, and countless others that preceded them, were indeed "blizzards," although it's unlikely that anyone labeled them as such, for in 1888, the term "blizzard" was a recent addition to the American lexicon and one that was virtually unknown in the East. It is a curious word, one without an identifiable Latinate root, for sure. It does have what poet Robert Frost once described as the "sound of sense," akin to a word that Lewis Carroll or Charles Dickens might mint to describe what their characters witnessed, although it is unlikely that either author ever experienced a blizzard.

H. L. Mencken reported a "single use of it" in the *Oxford Dictionary* in 1825, and lexicographer Allen Walker Read said it was used to denote a "violent blow" before it became associated with snowstorms.[4]

Read dated the earliest use of the term in the United States to 1870, when it appeared in an issue of the *Upper Des Moines* on at least three occasions. The phrase evidently gained popularity quickly. In the minutes of a baseball club meeting in Estherville, Iowa, that same year, the team decided on the name "Northern Blizzards," professing a "certain liking for it because it is at once startling, curious, and peculiarly suggestive of the furious and all victorious tempests which are experienced in this northwestern clime."[5] Dr. Stern noted that his daughter Meta's announcement on the morning of March 12, 1888, was the first time he had heard the word.

Today, the term is more technical than poetical. For a snowstorm to be declared a "blizzard," and thus so entered into the climate record, the government requires that certain criteria be observed at a "first order" official measuring station that is subject to its quality controls. The threshold is three hours of snow that is accompanied by three hours of sustained winds or frequent gusts of at least 35 miles per hour or faster or quarter-mile visibilities.[6] In 1888, the government had only recently established the U.S. Weather

Bureau, and the observation network still was a work in progress. The term "blizzard" had not yet entered the glossary. It certainly was not in the forecast language on March 11, nor was the conventional word "snow," and that was a profound humiliation for the weather bureau.

In the March 13 edition, the *New York Times* used the term. "There is no authentic record of a time since Manhattan Island was settled by white men, when the ordinary business of life was so completely brought to a standstill as it was yesterday," it said. "Due to the elemental fury of the day and night just passed," the *Times* advised readers that it could offer only a "somewhat partial and incomplete account" of the doings of the world.[7]

"It had a power of slinging the snow into doorways and packing it up against the doors; of sifting it through window frames of piling it up in high drifts at street corners," said the *Times*, "of twirling it into hard mounds around elevated station, such as most New Yorkers had never seen before. For the first time in their lives they knew what a Western blizzard was."

Officially and by any other measure what shrouded New York was a genuine blizzard and in the phrase of that Iowa baseball club, "all victorious." "New York helpless in a tornado of wind and snow which paralyzed all industry," the *Times* wrote, "isolated the city from the rest of the country, caused many accidents and great discomfort, and exposed it to many dangers."[8]

Blindingly heavy snow fell that Monday. Winds exceeded 50 miles per hour. "The air was full of strange flying objects, signs torn loose from shops, . . . awnings, frozen sparrows." The wind "caught up the dust and ash, swirled it about with icy snow, and flung the whole concoction into the faces of people and beasts." Some described it as being "like a barrage of bullets . . . like sand . . . like lashes of a whip . . . like flying glass."[9]

Temperatures plummeted from the low 50s late on March 11, to six above zero on March 12. Between 12:10 a.m. and 3:00 p.m., 15.5 inches of snow fell upon Manhattan, with winds gusting at almost 40 miles per hour.[10] The East River froze solidly. Most of the 200 deaths in the city, itself, were the result of hypothermia; some of those who had managed to get to work that morning—mill workers who failed to show up would be docked a day's pay come blizzard, tornado, or tidal wave—never made it home. To the great relief of his family, Dr. Stern was one of those who did survive it.

What made this one preeminent among all U.S. snowstorms was not only where it occurred—in the nation's largest city—but also when. This was the height of the industrial era. New York and the nation were in ascendancy, fast-forwarding into a technological revolution.

It was a heady age. Twelve years after Alexander Graham Bell invented the telephone, 10,000 New Yorkers had them, and nine years after Thomas

Edison founded his electric company, the streets were lined with 1,500 electric lights. When the Statue of Liberty was dedicated two years earlier, its torch was illuminated by electricity.[11]

The mountainous drifts of snow, the white snow-barricaded streets, and the magically whitened architectural details of buildings were captured in photography, further evidence of astounding progress and modern ingenuity. (Wilson Bentley captured a blizzard flake with a microphotograph.) But as striking as those street scenes were, something else in those images commands attention. Everywhere are the dense, homely, drooping wires, as though they were part of an artist's prepainting sketch lines of a street scene. The avenues appear as though they are tangled and bound in the wires that had connected Gotham with the world and vice versa. For 48 hours, New York and the world were disconnected. One could believe that the entire Industrial Revolution lay buried under the all-conquering snow.

"It was as if New York had been a burning candle upon which nature had clapped a snuffer, leaving nothing of the city's activity but a struggling ember," declared the *New York Sun*.

Almost all the electric lights in the city were rendered useless by the storm and most of the gas lamps extinguished. On the night of March 12, only snow illuminated the streets of Manhattan.[12]

The storm produced a blizzard of anecdotes, including one about a man who fell into a snow drift and cut his head on a rock-hard object. "A horse's hoof! That was a horse's hoof I fell against," he proclaimed to passersby. "Kicked by a dead horse, by cracky!"[13] And what a harbinger that tale was for the changes about to accelerate. In Springfield, Massachusetts, a man attempted to pick up a hat atop a snow pile, only to find that it belonged to a young girl who was desperately trying to tunnel her way out of the mass of snow with her hand.[14] The storm not only shut down schools, businesses, communications, and all means of transportation, but also forced the postponement of funerals. In Hartford, Connecticut, corpses were stored in frigid barns, and in New York, in snow banks.[15]

When the winds subsided on Tuesday, March 13, New Yorkers confronted an overwhelming challenge: overcoming nature's blockade. The major rail companies serving the city were equipped with plows, but they were effectively disabled. One train was overmatched by a 30-foot drift in the Bronx, causing a backup that stranded passengers for two days. It eventually was liberated with the help of a brigade of shovelers.

Shovelers were in demand as businesses and residents desperately sought hired hands to create paths to their doorways, a boon for the unemployed. For street-rail operators, finding shovelers to clear their tracks was about as hard as finding coins in a snow bank. Several of the companies opted for sleighs over street cars.[16]

The estimated two feet of snow that stopped New York City was an accumulation record that would stand for decades, but twice that amount fell to the north, in Albany and parts of New England, and 50-plus inches fell in Troy, New York. In no city, however, was the storm as deadly and debilitating as it was in Gotham.

Time to Get Serious

The blizzard was a deeply humbling experience for not only the forecasters, but also the city; the advances in communication and transportation that were catalysts of prosperity were rendered useless. The city learned that its battle plan for clearing the streets wasn't a plan at all. Garbage collection, already undependable, had to be put on hold indefinitely, a crisis throughout the city, particularly in the dense neighborhoods on the Lower East Side. The city gave the Department of Street Cleaning an emergency $25,000 appropriation to attack the blizzard's fallout, and the commissioner hired an "Army of the Shovel," like a Western sheriff forming a posse. It concentrated on the commercial cores of Manhattan, but the cleanup would require far more than just a posse of shovelers making Fifth Avenue safe for commerce. The challenges were immense and complex. The mangled overhead wiring had to be somehow repaired, the utility poles replaced, the rail lines cleared, the streets made navigable, and entrances to businesses made accessible. Shovelers loaded carts that deposited the snowy refuse in the East and Hudson Rivers, although some of the more creative laborers set fires that melted the snow banks, leading to gutter and basement flooding.[17]

The Blizzard of 1888 was an immense tragedy. Yet, it also was a gift to New York and other urban centers in the nation. With the complications attendant to progress and population growth, the storm underscored that snow-fighting was serious business that required more than the hiring of brigades of shovelers and random, unregulated plowing.

New York and other metropolitan areas eventually benefited from nature's punishing lesson. The blizzard and its aftermath argued eloquently for the need to prepare for snowstorms; accept the realities of how they affected life in rapidly developing urban environments; and recognize the need for responding to snow emergencies quickly, thoroughly, and coherently.

The blizzard provided fresh impetus to a radical concept: creating an underground transit system. But such a mammoth and expensive project would take several years to execute. Boston's would be completed before

New York's. In the years after the great blizzard, public transit remained an above-ground enterprise and one that was gaining steam or, more accurately, electricity.

When Horses Became Dinosaurs

That was the result of a development in—coincidentally—1888, in a city unaffected by the White Hurricane. It was a development that would intensify the war on snow and the demand for an escalation.

For decades, in New York and other cities, public transportation was driven literally by horse power. Animals hauled the carts, or cars that were attached to underground cables. Then, in the winter of 1888, a month before the blizzard, the world's first electric street railway system began operating in Richmond, Virginia. It was backed by New York financiers and developed by Frank Julian Sprague, an associate of Thomas Edison. Edison was focused on the lightbulb, but Sprague saw the potential for electrical power to drive the rails. Sprague, a native of Massachusetts, is credited with being a key figure in the electrification of Grand Central Station and was founder of a company that was the predecessor of General Electric.[18] Sprague's team created four-wheeled vehicles that were attached to overhead wires called trawlers, thus the term "trolley."

"The Richmond Union Passenger Railway stimulated a worldwide urban transportation revolution." That might be a mildly hyperbolic statement by the International Institute of Electrical and Electronics Engineers, of which Sprague was a former president.[19] But indisputably, the popularity of the electrical railway spread rapidly. Boston decided in the fall of 1888 that it needed to invest in such a system. Brooklyn signed on in 1890, and within seven years of the great blizzard, almost 900 electric railways were operating more than 11,000 miles of track nationwide. The trolleys not only were instruments of city growth, but also promoted and provided the cities with precious imports for the urban economies: workers who could now commute from neighboring communities.

"The cores of central cities grew and prospered as neighborhoods and streetcar suburbs developed along electric street railway lines," in the words of the electrical institute. "These lines served as an economic catalyst that directly linked consumer markets with marketplaces and workers with work places."[20] They also were the catalyst for the twilight of the urban horse. A generation after the blizzard, the cable cars and horse carts had become urban dinosaurs and all but disappeared from New York City streets.

If any year could be identified as the absolute terminus of the Pax Nivalis, it would be 1888. The occurrence of the blizzard and the development of a transit system that would make cities more accessible to inter- and intracity commuters suggest a certain symmetry. The rise of the motorized vehicle yielded abundant harvests of opportunity and convenience—and challenges. The rails were prone to icing, and the cold metal was the ideal environment for freezing and re-freezing snowmelt, a peril to the railed vehicles. And the trolleys were merely a prequel for the complications to come with the motor age, which would provide jolts of intensity to the enmity between snow and an ambitious society.

As historian Bernard Mergen noted, more and more snow was viewed as a hazard that had to be pushed out of the way.

A General Joins the War

"The question of snow removal has always been of the most vexatious problems confronting various administrations."[21] Those words could well have been written in the 21st century, germane to any of the thousands of U.S. cities and towns affected by snowfalls. In this particular case, they were contained in a 19th-century report presented to the New York City mayor's office by one George F. Waring. It was written by a subordinate, "snow inspector" H. S. Stidham, a 42-page report packed with data, straight talk, and maps. It was a postblizzard snow-fighting manifesto.

If the war on snow had a George Washington, a likely candidate would be Waring, a Civil War veteran. He was appointed New York City's chief of streets and sanitation by Police Commissioner Theodore Roosevelt. Waring, a highly respected civil engineer, was a flamboyant figure who ordered his employees to dress in white and was known to lead his "White Wings" along the streets of New York, himself dressed in white and on horseback, presenting the city and the nation with a very different image of the sanitation worker in the 1890s.

By the time he took command, the city had instituted some upgrades in its snow-fighting plans. In 1892, Waring's predecessor identified areas of Broadway and other business arteries that would be snow-clearing priorities and ordered his street cleaners to suspend all other activities during storms and focus on keeping the city moving; however, neither his cleaners nor the immigrant shovelers he hired bothered to remove snow from the streets around those teeming tenements on the Lower East Side, where uncollected garbage was allowed to rot, exposing hundreds of thousands to potentially serious health hazards and might have resulted in deaths, according to a report commissioned by Waring. Some frustrated residents who were fed up with waiting for garbage trucks dumped their refuse onto snow piles, where it could mutate into a toxic soup.

A further complication to the removal program was organized labor's success at getting a law passed that restricted municipal employment to U.S. citizens; that eliminated the enlistment of inexpensive immigrant labor. When Waring took over, to get around that problem he pursued an option that today is standard municipal practice: He outsourced, contracting with independent operators to supply carts, shovelers, and manpower.[22]

In the winter of 1896–1897, 144.4 street miles were targeted for snow clearance, almost a sevenfold increase from just a few winters before. That did not include the efforts of the rail companies. Waring wisely had negotiated an agreement that accorded them responsibility for clearing not only their tracks, but also the streets on which they operated, from curb to curb. He expanded the department's snow-fighting territory, adding those crowded areas east of the Bowery and south of Houston Street. For the first time, blocks near schoolhouses were added to the list.

Shortly before he died, he presented to the administration a comprehensive manual that included a treatise on snow removal written by his "snow inspector." It noted that after the Blizzard of 1888, under the supervision of Waring's predecessor, the city and various contractors had carted away "40,542 loads of snow." By comparison—five times that amount—more than 200,000 loads of snow—had been removed in each storm during the winter of 1896–1897, in a season in which the biggest snowfall was well less than half that of the blizzard. That did not include the amounts removed by the rail companies, whose efforts saved the city an estimated $3.5 million in today's dollars in the winter of 1896–1897, or $300,000 more than the department had spent annually from 1882 to 1894.[23]

While plows were pushing aside snow on some streets, New York City's emphasis was on removal. In the Waring report, plowing is hardly mentioned, appearing near the end in the "experiments" section, which included references to the short-lived and impractical flirtation with snow-melting devices. Given the frequency of snowfalls and the cubic volume of snow, the only realistic option for melting was that object that rose in the east every morning.

"We're Tougher"

Yet, by New York state standards, the city was in the banana belt. Manhattan island is a finger of land that climatologically has far more in common with New Jersey than the rest of the Empire State. Manhattan parallels the North Jersey coast; the George Washington Bridge links the northern end of the island with Fort Lee. Snow Belt residents are prone to find New Yorkers' complaints about snow amusing, if not annoying. In Syracuse, Buffalo, and Rochester, amounts are triple and quadruple those of Gotham. They rank among the snowiest populated areas of the nation, and in the late 19th century, they were expanding rapidly, their economic fortunes buoyed by the Erie Canal. Their combined populations doubled, to about 450,000, between 1870 and 1890. As they grew, so did their winter issues.

With such influxes of new residents, hundreds of thousands of newcomers were having their first encounters with the lake-effect snows that were winter constants on the lea sides of those Ice Age leftovers known as the Great Lakes. They were discovering what the veterans well knew: The winters were far snowier and more challenging than they were to the south and east—and not all that far. Ithaca, for example, is only about 40 miles from Syracuse and receives roughly half the snow—or about as much as Denver. One Syracuse ex-patriot who moved to Ithaca said to me, "We just don't get the snow down here." Yes, she was complaining.

The snows did not discourage development and commerce, and arguably the challenge of adaptation contributed to economic growth. Rochester became the headquarters for Union Telegraph, whose silence provided the first clues about the severity of the Blizzard of 1888 that was shutting down the New York state banana belt; the Bausch and Lomb optical company, so instrumental to snowflake research; and, of course, Eastman Kodak, which would acquaint the masses with the new marvels of photography. Buffalo, Rochester, and Syracuse learned to coexist with seasonal snow totals that far dwarfed those of Philadelphia and New York City, and even those of Albany, Boston, Detroit, Chicago, and Minneapolis. I once asked a Syracuse school superintendent how the natives managed to conduct the affairs of life through all that snow, and his answer was simple: "They are rougher and tougher."[24]

But even the lake-effect cities had their limits. They were among those embracing the electrified transit systems, and in 1893, some of those rail companies would have their first serious encounter with the extremes of winter. Three feet of snow landed on Buffalo, and the city had to send in the plow horses to bail out its electrified marvel. Fortunately, the city still had six horses available to pull snowplows and keep at least a few lines open. To haul away the plow walls, it hired shovelers to clear the downtown area. Similarly, Rochester had to go retro and resort to horse plows, but it was about to take a significant step toward the future of snow-fighting, and other cities would follow.

That same year, the inventive Charles W. Ruggles secured a patent for an electric rotary plow that became an effective snow fighter on the rails. Ruggles was a Lake Ontario ship captain and likely was inspired by what he had seen of vessels' paddle wheels and propellers. His electrified plow car was equipped with fan blades that chopped up the snow like a modern-day food processor, and blades would hurl the snow to the left or right.[25]

Two winters later, equipped with the Ruggles plows, Rochester's rails were able to keep operating as the rotary blades broke through the drifts. Ruggles signed over the manufacturing rights to the Peckham Motor Truck and Wheel Co., in Kingston, New York, and the plows became staples in U.S. and Canadian cities. The plows received rave reviews.

"I confidently assert that it will do more work in an hour than 50 men with shovels in a day," wrote the head of the Niagara Falls and Suspension Bridge Railway Co. "Your machines have removed snow from four inches to five feet in depth without the aid of a shovel," commented the head of the Rochester Electric Railway.[26]

More properly, the writer probably should have used the verb "moved" rather than "removed." Plows push snow aside, not remove it. Removal remained a major headache for snow cities, particularly in the heart of winter when the sun angles were low and cold spells tended to endure. The sun gains ever-increasing power in late winter, and as burdensome as the removal operation was after the Blizzard of 1888 in New York, the March sun was a powerful ally. February is by far the month when the sun makes its biggest gains in radiative power in the midlatitudes in the Northern Hemisphere, and by March 1 the amount of solar radiation reaching the midlatitudes is double what it was at the winter solstice, according to Drexel University atmospheric scientist Fred House.[27]

The price tag of removal varied with not only snow amounts, but also timing. For example, in 1900, Rochester experienced an unprecedented late-winter blitz, a three-day storm that ended on March 2, leaving 43.2 inches of snow. Yet, the removal costs that winter were only a third of what they were two seasons later, when the snow totals were higher. It so happened, however, that in the later season, the storm sequence had begun in January, when the sun wasn't much help.[28]

Removal remained a perennial issue in every city affected by snow. While the rails were getting better at pushing it aside, the cities were left to figure out how to keep streets and sidewalks passable and provide as much access as possible to their downtowns and commercial centers.

A disaster in the Rochester business district in February 1904 underscored the complications. A fire broke out during a snowstorm, and in addition to smoke and flames, firefighters had to battle plowed snow piles blocking crucial access streets. The mayor ordered the city engineer to come up with a better snow-fighting plan. Following New York City's lead, Rochester expanded its program significantly, extending sidewalk clearing from 12 to 83 streets.

Other snow-challenged cities opted for a variety of strategies—and nonstrategies. Although snowier, Syracuse settled for concentrating on selected crosswalks and sidewalks, leaving it to the trolley company to plow the rails.

To the west, the metropolitan behemoth of Chicago, whose population had reached 1.7 million by 1900, put ward superintendents in charge of snow removal. Contributing to a rich tradition, the operation became the subject of a scandal in the First Ward, the site of the Loop, shortly before Christmas 1907. The ward chief hired more than 300 men to shovel snow into horse-drawn wagons that teamsters would drive along Van Buren Street and dump the snow into Lake Michigan. The workers were paid by the cart load. To bump up the numbers and thus pad the payroll, the teamsters would dump only part of the loads, making the refills that much easier and increasing the numbers of trips. Unfortunately, a witness caught them and reported the cheating to the outraged mayor.

Snow removal was costing Chicago about $700,000 a year in today's dollars, and January 1910 was so snowy that the fund was depleted, forcing the council to suspend the service.[29] What was becoming more evident in the 20th century was the reality that the costs of removal were mounting, and the ad hoc approaches were no longer workable.

Relativity

The impacts of snow on a given city always have been relative to the snow amounts and frequencies to which a city is accustomed. Two inches of snow in Atlanta, for example, could be more debilitating to civic life than two feet in Syracuse. We visited Syracuse one weekend some years back. Ten inches of snow had fallen in two days. In Philadelphia, that would be a showstopper, and the local TV stations might have gone all snow, all the time. In Syracuse, no one seemed to care; it might as well have been partly cloudy. At intersections, snow was piled so high it was rather disconcerting that we couldn't see the cross traffic; obviously these folks were used to it. I'll never forget the story that Syracuse's former chief snow fighter, Phil Wright, told me about the March 1993 blizzard. At the time I interviewed him, Wright and the city highway crews had just survived a crisis. Syracuse had been hit with a lake-effect snow blitz in which 10 inches piled up in just two hours.

It shut down the city—for three whole hours. As for what 10 inches in two hours would do to Philadelphia, I don't think I'd want to know.

But Wright said that the Blizzard of 1993 truly did stop him. That would have been understandable. A total of 43 inches of snow accumulated on Syracuse during a storm that generated gusts faster than 45 miles per hour. But it so happened that Wright wasn't in Syracuse. He was in New York City for a convention and stranded at his hotel. Officially, 12 inches was measured in New York.[30]

While Syracuse residents are entitled to a certain disdain for the snow-whining emanating from the Northeast corridor, those occasional outsize storms have been especially disruptive in the coastal-plain cities, including Philadelphia, creamed by a surprise 21-inch snowstorm on Christmas Eve 1909.

Thus, it is not all that surprising that the city that hosted the Constitutional Convention would be the site of the nation's first convention on snow removal. In March 1914, the city's chief of public works, Morris L. Cooke, summoned his counterparts from Washington, Boston, Pittsburgh, and elsewhere to meet at Philadelphia's City Hall, a Second Empire landmark whose architectural details were especially decorative after a snowfall, to discuss their wintry issues and what to do about them.

They gathered the following month and addressed a wide range of snow-fighting methods, from the promise of snow-melting machines to the one that had stirred opposition in the 19th century: salt.

The use of salt as a melting agent was popular in European cities, including Paris and Liverpool, the report noted. Salt transformed the frozen material to a slurry, then to liquid that was flushed down storm drains. But in the United States, salt was for sprinkling on food, not streets. New Yorkers complained about the splashings of the slushy residue upon their clothes. Given the accumulations of trash and garbage, not to mention horse manure, upon which the snow had landed, such objections were understandable. The horses themselves were salt's victims, according to animal-rights activists. The American Society for the Prevention of Cruelty to Animals raised a "serious objection" to the spreading of salt on city streets, and several cities outlawed it, notably New York. (The *Times* reported that in one instance, the society had to destroy horses that had suffered "hoof-root" caused by salt, which "dries up the horses' legs, and the exposure to the weather cracks and scars them.")[31]

Yet, the officials gathered in Philadelphia advocated giving salt a chance.

"It is questionable whether the use of salt has been given a fair trial in this country, and there is little doubt that it would be useful in light snowstorms," the delegates concluded. The overriding limitation of salt, of course, was that it was quite insufficient for dissolving five-foot drifts. Paris and Liverpool had no experience with North American mega-storms.

Salt was not going to work on its own. Drawing on the experiences of New York, Boston, Philadelphia, and Scranton, along with the Pennsylvania Railroad and the Public Service Railway of New Jersey, the delegates concluded that the only way to fight snow was to develop comprehensive citywide battle plans and that they should be formulated well before the arrival of the first wintry storms. They recommended that removal work "should commence as soon as the snow has covered the pavements and the indications point to the storm continuing, and should be carried on continuously." They also suggested that city officials instruct their constituents on how to clear their own properties. They should clear paths, and rather than dump the snow into the streets, it "should be left on the sidewalk near the curb line to be later removed by the city when opportunity presents itself."[32]

After the conference, cities took action. Philadelphia, the host city, targeted 20 miles of streets for snow clearance and assigned 200 municipal employees, with the option of hiring added help, for snow clearance. Each of Philadelphia's 19 districts had a designated dumping area. Boston commissioned its 1,000-man street-cleaning force to handle moderate snowfalls with the option of hiring more laborers for heavier snows. Chicago also enlisted its street cleaners.[33]

The timing of the Philadelphia conference was fortuitous: The report of the proceedings took note of a "general increase in motor traffic." That was true, but in that regard, the delegates hadn't seen anything yet.

The Road Ahead

I will build a motor car for the great multitude. It will be large enough for the family but small enough for the individual to run and care for. It will be constructed of the best materials, by the best men to be hired, after the simplest designs that modern engineering can devise. But it will be so low in price that no man making a good salary will be unable to own one—and enjoy with his family the blessing of hours of pleasure in God's great open spaces.—Henry Ford[34]

In 1900, 8,000 motor vehicles were registered in the United States. By 1914, the number had jumped to a little less than 1.8 million, and that total increased fivefold before the decade was over.[35] Henry Ford and the mass production assembly-line revolution that made cars affordable for the people who made them were putting Americans on wheels—and yet another collision course with nature.

The arrival of the motor age did afford cities one advantage in that they now were able to acquire a new generation of motorized plow vehicles. By the end of the winter of 1916–1917, New York had a fleet of 96 of them and was able to clear almost 1,000 miles of streets. Chicago's Mercury Manufacturing Co. and the Good Roads Manufacturing Co., outside Philadelphia, made plows for several cities, including Buffalo and Syracuse. Foreshadowing what would become a popular adaptation, Newark used its road-building shovel to clear streets in the winter of 1917–1918, one of the harshest on record in the Midwest, Ohio Valley, and Northeast. New York would order 100 five-ton tractors with treads like Army tanks to plow through snow piles and move gravel and dirt in the summer. Such companies as Caterpillar and Mack were marketing snow-removal equipment that could be used for road building and maintenance in the warm seasons.[36]

The motorized cities, becoming evermore vulnerable to the impacts of winter storms, found themselves in an arms race against nature. Congress authorized Washington, DC, to acquire plows after the tragic "Knickerbocker Storm" of January 1922. Two feet of snow shut down the nation's capital, and the weight of the frozen mass caused the collapse of the roof of the Knickerbocker Theater, in the city's Adams Morgan section, killing 99 people.

Elsewhere, the movement toward organized, officially directed snow-fighting was gaining momentum. Albany and other towns in upstate New York and New England took on responsibility for clearing snow and ice from their business districts, although in some towns it was left to chambers of commerce to raise the money for equipment.[37]

Complicating the cleanups and adding all the more urgency to them was the staggering growth of automotive traffic. An extensive road network, which served even sparsely populated areas of the nation, was in place well before the motor age. In 1900, only 4 percent of the roads were paved, but after World War I and the creation of the Federal Aid Highway Program, that would change dramatically, along with the volume of traffic. The numbers of registered motor vehicles doubled in just four years, to 15.1 million, from 1919 to 1923. To manage maintenance of state roads outside the cities, Congress decreed that states create highway departments. Those departments would become heavily involved in blunting the effects of winter storms. Following Connecticut's lead, Pennsylvania, New Jersey, Massachusetts, Colorado, and Michigan took on plowing and sanding chores for their state roads.[38]

The asphalting of the United States and the associated boom in motor traffic put people on not only wheels, but also, in winter, thin ice. More commuters were opting for cars, and buses were replacing trolley lines. The meeting of wheels and pavement created new hazards. Plows, when they weren't inhibited by stranded motor vehicles, could make streets passable. Foreshadowing what would become standard practice, in 1926 an engineer with the Michigan highway department used a sophisticated model to make the case that straight-blade plows attached to trucks would be the most efficient way to move snow off roads. As did the delegates at the Philadelphia conference 12 years earlier, he recommended commencing plowing operations at the beginning of a storm.[39]

But the action of traffic could make streets all the more perilous as tires compressed snow and ice to form dangerous frictionless surfaces. Thus, even nuisance winter storms with minor accumulations were becoming problematical in the 1920s. More motor-driven wheels meant more sensitivity to wintertime storms. Car crashes were on the upswing. The invention of tire chains in 1904 was an important contribution to driver safety; however, chains were punishing to road surfaces, and they weren't built for higher-speed commuting. Cities throughout the country, and not just those prone to frequent heavy snows, were forced to draw up plans for winter's annual challenges, to develop ways to make roads passable and safer. "The automobile today has passed from being a pleasure vehicle to a public necessity," declared the Automobile Club of Berkshire County, Massachusetts, and it was imperative that roads "be kept open to traffic throughout the year."[40]

An era of road treatment was underway, and eventually the nation would turn to that prosaic and previously disdained ingredient. As the delegates to the Philadelphia conference had recommended, they were giving salt a chance.

Road crews were spreading sand and cinders atop plowed snow and ice to improve traction, but those materials did nothing to promote melting. The complicated career of snow and ice was and is particularly problematic on paved surfaces. The freezing, melting, and refreezing processes were constant threats to motorists. The ideal solution would be to find an effective melting agent that would transform the frozen cover to liquid that would sublimate and/or evaporate and leave the paving dry. A breakthrough came in 1926, when Cleveland Railway Co. experimented with calcium chloride, finding it effective in melting thick layers of ice. More and more cities were indeed opting for a melting agent—not calcium chloride, but an ubiquitous chemical cousin.

High-Sodium, and Major Arteries

Beneath the shores of Cleveland, 1,800 feet below Lake Erie, is an environ-ment that evokes a setting for a horror movie; the rutted roads are bracketed by slush-colored walls and adorned with ghostly, billowing curtains. This might be the safest place to be during a Great Lakes blizzard. Here the rem-nants of ancient seas that geologists date to the Silurian Age have bestowed upon the nation a vast repository of sodium chloride, or salt. Rival giants like Cargill and Morton own others. Delivery is another matter altogether, but supply never has been—and likely never will be—an issue. The U.S. Bureau of Mines has estimated that since the late 18th century, the nation has ex-tracted 2 billion tons of salt from its mines.[41] Syracuse, the snowiest large city in the country, is known as the "Salt City." That is not related to road salt, however. In the 17th century, Jesuit missionaries discovered salt deposits at the base of what is now called Lake Onondaga, about 15 miles south of town, and the region once had a thriving commercial salt industry. Mined salt is still important in Syracuse; it relies on salt for its melter these days.

As much as Syracuse needs salt in winter, it is not a threat to the nation's supplies, despite its prodigious snow totals. Given that the country sits atop about 60 trillion tons of salt—enough to fill 20 million 50-story skyscrap-ers—and that about 15 million tons are used annually, road departments in Syracuse and elsewhere should be good for the next 4 million years or so. And given the quantities, it is not surprising that sodium chloride has always had one huge advantage over calcium chloride: It is way cheaper.

It also helps that it works, at least with temperatures above 15. The ice-repelling properties of salt were long known, and Gabriel Fahrenheit is cred-ited with documenting in the 18th century that it lowered the temperature at which water would freeze.

A study by Marquette University found that the treating of roads with salt reduced injury-causing accidents sevenfold.[42]

Cleveland, Buffalo, and Detroit were among the cities that saw the advan-tages of salt and exploited them in the 1930s. Springfield, Massachusetts, and Schenectady, New York, joined them.

Salt, a major flop on Broadway in the 19th century, made dramatic in-roads in the 20th century; however, its environmental impacts remained a source of consternation. Samuel N. Baxter, a legendary arborist with the Philadelphia park system's tree commission and a local legend, complained that ice-cream salt dumped from delivery wagons was killing trees.[43] Salt's career as an environmental issue has only prospered in the 21st century with growing evidence that it is a significant threat to vegetation and water supplies in areas that do annual battle with snow and ice. That would be most of the country.

Saltwater

Traces of chloride in streams and rivers are common. What was found in a stream running through Philadelphia's prestigious and highly developed Main Line wasn't. Water samples taken in May 2019 showed chloride concentrations higher than 800 parts per million, more than triple the levels potentially toxic to aquatic life.[44] Chloride and sodium ions separate when they dissolve in water. Thus, chloride is an excellent proxy measure of salt.[45] Unlike Syracuse, Philadelphia is nowhere near any source of natural salt deposits, pointed out Samantha Briggs, who runs the Izaak Walton Conservation League in Gaithersburg, Maryland, which has organized the national water-testing network.[46]

Nationwide, chloride concentrations in freshwater have jumped dramatically since the 1940s, coinciding with the dramatic leaps in winter salting. The findings in the Philadelphia region, where the volunteer monitoring network is particularly aggressive, have been especially disturbing. Darby Creek courses through the Main Line, and all the major and no-name streams in the area drain into the Delaware River, an important source of local drinking water.

It can take a decade or more to wash chloride out of freshwater. Ironically, Briggs notes, chloride concentrations in streams can spike during heat waves as water evaporates in the baking sun.[47] Not surprisingly, salt has found itself in environmental activists' crosshairs, the ground-level equivalent of greenhouse gases in the atmosphere.

On a more prosaic and short-term level, no one who has spent any time in a snowy venue needs to be told about salt's effects on vehicles, shoes, and floors. By one estimate, corrosion damage to automobiles, metal rods in bridges, and roads, along with measures to protect them, costs the nation almost $20 billion a year.[48]

On a practical level, the cost of traffic accidents and their effect on lives are immeasurable, and any movement to ban the use of salt entirely has gained no traction. But spurred by environmentalists, road departments throughout the country have been looking for ways to reduce salt use.

Grapes, Vodka, Cheese, and the Anti-Icing Revolution

One strategy that has become increasingly popular is the process of "brining," which has wrought a modest revolution in snow-and-ice fighting. It has not only changed how roads are treated, but also opened up new possibilities for exotic snow-and-ice-fighting recipes that go beyond salt.

Throughout the country, when snow or ice is in the forecast, road departments have been spreading a liquid mixture of water and salt, in some cases with dashes of magnesium chloride, on their networks of streets and highways before anything has even fallen from the skies. Motorists and residents now are quite familiar with those brine lines, which resemble plow rows on a farmer's field. Where I live, I have seen roads brined so heavily that I have half-expected something to start growing out there. One researcher, Washington State University's Xianming Shi, suggested that the image might be appropriate; he believes the remedies to salt might be growing somewhere near all those brined roads. No, he says, the idea is not in the least far-fetched, thanks to the advances in brining.[49]

For salt to work as a melting agent, it has to dissolve into a brine. When dry salt is applied to a road, the precipitation is the source of the liquid that dissolves the pellets, and the resulting brine lowers the freezing point of the water.

When prewetted brine is applied before precipitation falls it adheres to surfaces, rather than bouncing off like rock-salt pellets. It deters falling snow and ice from bonding to the road surface. It is an "anti-icing" strategy, rather than "deicing." It effectively reduces the need for salt and limits salt waste. "Brining is a better solution in itself," said Samantha Briggs.[50]

And Shi is convinced that with additives made from local sources it can be an even better solution by further depressing the freezing point of water and the need for salt. With grant money from the Alaska Department of Transportation, Shi and a team of researchers gained a measure of celebrity in 2012, by cooking up an effective ice-repelling recipe that included leftover barley residue from Alaska's distilleries.[51]

Highway departments have experimented with a variety of additives. Colorado's has been adding a corrosion inhibitor to the slurry, and beet juice has gained a following. New Jersey's transportation department says beet juice has been effective because it acts as a low-grade epoxy for its anti-icing materials.[52]

Shi said he has found something better. In Washington, he began looking at a local agricultural product, the concord grape. In two years of laboratory testing, he documented that a deicing compound with a grape extract outperformed standard salt-brine solutions, reducing the freezing temperature to 11 below zero Fahrenheit.[53]

In often-frigid northwestern Wisconsin, Polk County highway chief Moe Norby said he harvested outstanding results from the state's signature agricultural product—cheese. He was trying to find ways to cut back on salt and sand when an idea condensed: Could he use a waste product, a local dairy's salt brine, on the county's roads? He visited the facility and took home two 12-ounce containers of the brine, which he left outside. It so happened that on consecutive nights the temperatures plummeted to 21 below zero. The brine didn't freeze. He was convinced.

The brine solution was strained and pumped into trucks that ferried it to the plant, where it was filtered to remove any solid whey residue. Using brine that the factory was going to have to dispose of anyway, he was able to cut back on salt expenditures by 30 percent. And, no, the cheese did not raise a stink.

Unfortunately, the county stopped using it after the winter of 2018–1919. It wasn't that it wasn't working; the dairy was bought by a bigger company and closed. Norby hasn't been able to find another partner within easy driving distance.[54]

"Brining works well under certain conditions," said Ann Fordock, the deputy streets commissioner in the nation's urban snow capital, Syracuse. But not everyone is sold on it. It has been linked to brake-line damage on cars and trucks, and it's ineffective if rain falls before it snows, a common occurrence in some Mid-Atlantic and Northeast cities.

Antecedent rain hasn't been a factor in Syracuse's decisions on when to brine. First of all, it usually doesn't rain there before it snows, and second, the city never uses brine, period. Fordock said she has been trying to persuade the veteran employees to give it a shot but so far unsuccessfully.[55] As of the winter of 2019–2020, the city was sticking with traditional salting.

Salt overwhelmingly remains the material of choice for brine mixtures, which typically are about one-quarter salt. Shi acknowledges that while salt will be a staple indefinitely, road departments should be looking at alternative additives, even if they drive up costs.

"By buying the more expensive product you save money in the long run," he said. "The hidden costs are not fully integrated into that decision-making."[56]

Samantha Briggs, however, said the nation should proceed with caution. The additives are so new that no one has assessed what effects they might have on the environment.

For now, salt rules.

Our Salt Dependency

Today along highways in the East that have become winter battlegrounds, you are likely to see those teepee-shaped salt domes virtually vomiting an excess of the workingman's wonder-melter. That excess is a byproduct of the winter of 1993–1994, unprecedented for ice in the Mid-Atlantic, snows in the Northeast and Midwest, prolonged cold that hasn't been matched since, and a puzzling shortage of one of the world's most plentiful raw materials.

Nothing happened to the supplies. The world still had enough to fill those 20 million 50-story urban skyscrapers. The problem was getting it to the places that needed it. By mid-January, it was clear that road departments had not ordered enough for a winter of this severity. For example, the Pennsylvania Department of Transportation had calculated its needs based on what it had used in the previous five winters. Save for that March 1993 storm, the winters had been certifiably wimpy. On Martin Luther King Day 1994, a major snowstorm socked the Northeast, and the very worst of a series of ice storms lacquered parts of the Mid-Atlantic. In the Philadelphia region, commuters were stuck on a major highway for eight hours, forced to use empty coffee cups for urinals.

The problem had to do with the complicated salt-delivery chain and the fact that highway conditions were so consistently perilous that the delivery trucks couldn't get to the places that needed what suddenly had become a precious commodity. When the trucks finally showed up, it was akin to the arrival of the calvary. That winter underscored the value of salt, and, yes, it corroded metal, made for unsightly slush that splashed on clothes, and got washed into streams, but damned if it didn't keep traffic moving. From that winter onward, highway departments resolved that they would not get caught short again. They haven't; typically, the Pennsylvania transportation department has enough on hand to get through two normal winters.

The winter of 1993–1994 was a defining season in the evolution of salt dependency. Before the 1940s, salt, itself, was used primarily as an additive to keep sand piles from freezing and becoming unspreadable, rock-hard clumps; however, as America's roads became evermore crowded with motor vehicles in the years before World War II, the hostile attitudes toward applying salt directly to paved surfaces were evaporating.[57]

New Hampshire, which throughout the years has launched and deflated presidential candidates, made a choice in the winter of 1941–1942 that forever boosted the selection of sodium chloride as a weapon of choice in the war on snow. It went all-in, relative to the times, for salt. The White Mountains of New Hampshire are spectacularly beautiful. The Greens of its neighbor to the west are more celebrated for scenery and winter sports, and that is understandable, but if I had to choose between the two, I'd take the Whites and those brooding snowcapped peaks. In the winter, from a distance, you can see how they earned their name. Mount Washington always reminds me of a cold old man wearing a white scarf and exhaling vapor.

The New Hampshire terrain also can be challenging, with steep climbs and descents that can put such stress on the brakes that you can smell the rubber. At the lower elevations, it snows—a lot. It is in the way of those clipper systems, and lest we forget, New Hampshire also has an Atlantic beach, and the state gets walloped by those nor'easters. Otherwise, it happens to be one of the nation's coldest states, being one of its northernmost. Maine, to the east, deprives most of the state from modifying Atlantic breezes, while not shielding it from invasions of Atlantic moisture.

Thus, the choice New Hampshire made in the winter of 1941–1942, to officially sanction the use of salt, wasn't all that surprising. That first winter, it applied a mere 5,000 tons.

That was a hors d'oeuvre. By 2015, the total had jumped 40-fold, to 200,000 tons.[58]

In the postwar years, snow-and-ice-free roadways became an expectation, if not an entitlement. The primary vehicle for meeting that demand was salt, and the increases in its use became exponential. In the 1950s winters, salt rivaled rock 'n' roll for popularity, and the sounds of tire chains crunching and grinding atop salt-and-cindered roads were the equivalent of urban jingle bells.

The Eisenhower administration was a boon for the salt industry. The nation undertook an unprecedented highway-building frenzy that accelerated the transformation of cities from urban centers to integrated city-states. Split-levels, ranchers, and Cape Cods were sprouting up in the towns outside the city limits, doubling and tripling their populations. Tens of millions of residents in those "bedroom communities" were commuting to jobs in the city centers. They had no time to stop for snow.

Snow had outgrown its career as "refuse," to use Mergen's term; it was an obstacle, a threat to regional economies. States, cities, and suburbs, evermore organized and equipped, were waging all-out battles against snow and ice in ways that no one could have envisioned in 1888.

The new realities and the blacktop imperatives coincided with, and no doubt accelerated, revolutionary advances in the art of weather forecasting. Governments, businesses, and commuters needed to know what was coming.

The humiliating ambush of the Blizzard of 1888 was unacceptable. Not that they would end, although meteorologists and physicists were gaining new understanding of why blizzards happened and why the weather on the western side of the Atlantic was so much different from what was occurring on the other side of the pond.

CHAPTER FIVE

Mighty Streams

The Conspirators of Blizzards

The unraveling of the mysteries behind the frequently violent weather of the United States begins more or less in the Far East. To borrow a phrase from naturalist Loren Eisley, it has been an immense journey, and one that looks to continue in perpetuity. It took decades and a world war for the clues to fall into place, but one significant—and tragic—fragment of evidence appeared in the woods of southwestern Oregon on a May day in 1945.

That day a stranger in a U.S. Navy uniform burst into the tiny switchboard room bearing clear and urgent orders. Keep quiet, he told the shocked teenage telephone operator. Stay put. Sixteen-year-old Cora Conner was scared to death.

It was the first Saturday of the month, a day that had begun so promisingly in the tiny town of Bly, Oregon, just north of the California border. Only two hours before, five young friends had stopped by the Conner household. They tried to entice Cora and her sister to exploit the splendidly sunny skies and long-overdue warmth, and join them at a Sunday-school picnic in the Fremont Forest, a short drive out of town. Cora was aware that she was being recruited as an accessory, albeit a useful one. One of the boys had a crush on her sister; the boy understood enough of romantic politics to invite both of them. Cora wasn't offended.

The girls did not have the chance to answer. Mrs. Conner said the picnic was quite out of the question for Cora, since it was her turn to work the jacks and cords at the switchboard the family operated under a contract with the phone company. If Cora couldn't go, likewise it would be quite out of the question for the sister.

It would be 45 years before Cora Conner could talk to anyone about that work shift. First, a huge man from the Forest Service had taken over the switchboard room, making a frantic call to the nearby naval base. Then the U.S. Navy officer arrived with his stern and chilling orders.[1]

Something unimaginably horrible had happened in the woods, a tragic event linked to a phenomenon that would become the focus of the atmospheric detective story that would empty into the riddles of why the United States was ambushed by cosmic snowstorms and what had chilled and shocked the first European settlers. One of the most significant advances in the history of meteorology had been hijacked for a horrific purpose.

And how horrible it was. Cora's five friends were dead, she learned that morning. She was not to leave the room, not even to go through the door that led to the rest of the house. She was not to speak with anyone, not even the mother who had saved her life by forcing her to stay home.

Outside, an agitated posse of townsfolk was swelling. People pounded on the door and the window glass, yelling, screaming, their familiar faces reddened. Ignore them, the man told Cora. The hours passed. At dinnertime, Mrs. Conner slid a plate of food through a slot in the door leading to the switchboard room. Outside, the ranks of the shocked and angry kept their vigil through the descending darkness. Along the streets, citizens of Bly were gathering in small groups, speculating on what had happened in the forest and why somber-looking men in uniforms were appearing in the town.

With each overheard phone conversation, Cora would learn more about the nightmare in the woods. She was evermore distracted and frightened by the banging and pounding of the people outside. The loudest, angriest, and most incessant of them all was a beefy man in a plaid flannel shirt and lumberjack hands. She tried hard not to look at him. She knew him well. His name was Ed Patzke. He wanted to know what in the living hell was going on. His brother and sister were dead.

At the end of that endless day, they would embrace, and in Cora's words, "We cried, and cried."

Cora Conner was among the last to see 14-year-old Dick and 13-year-old Joan Patzke alive. The Patzkes, Edward Engen and Jay Gifford, both 13, and Sherman Shoemaker, 11, were the ones who came to the Conner household with the picnic invitation. Their chaperones and hosts were the Rev. Archie Mitchell, pastor of the Christian Alliance Church, and his wife, Elsye.[2]

The Mitchells were new to town, and this would be a get-acquainted outing. Archie Mitchell, tall and angular with wire-rim glasses, had just turned

27. Elsye was five months pregnant. In their wedding picture, they made a striking couple. Even in a wedding gown Elsye's expression suggested she did not like being photographed; however, her veiled discomfort softened her serious face.

Elsye Mitchell almost missed the picnic. Although she had made a chocolate cake on the day before the outing, she was not feeling well. On Friday night she had just about decided it would be wiser to stay home and take it easy on Saturday. Then she thought about the children. They would be disappointed, for certain. Dick Patzke, almost 15, kept bugging her about why she was so sick so often. Elsye confided to her sister that she was prepared to give him a lecture on the facts of life, "but thought I wouldn't," she said. He would learn them someday.[3]

On Saturday morning the sun radiated magnificently above the snow-crested Cascades and melted her resistance. She could see that Archie was anxious to mine this good fortune. The children would be even more excited than Archie. She watched him load the fishing gear in the sedan. They would be shopping for a venue near a creek. Elsye reevaluated her condition. She couldn't miss the opportunity to luxuriate in the day, savor the woods, and bond with the young people of their new congregation. She decided to go.

She soon questioned her judgment. It was a challenging and nauseating ride for her. It was a mere four miles to the park entrance, but the forest road was rough and weathered with the ravages of winter. As the sedan crept and wobbled toward Gearhart Mountain, the rocky, muddied roadway jounced the car. After five-plus torturous miles, Elsye was becoming carsick. When her husband stopped the sedan, however, it wasn't at her request. The party encountered a pickup truck and three workmen who were extricating a Forest Service road grader from the May mud.

Richard "Jumbo" Barnhouse, the road crew foreman, informed Archie that the road was impassable. It was just as well; Elsye needed to get out of the car. Archie observed that what he knew to be Leonard Creek was a mere 50 yards away. He asked Barnhouse about the fishing potential. Barnhouse said the creek might be too full of muck. No matter, they could go no farther anyway. Elsye rushed off to explore the creek with the children, although she couldn't keep up with them as they raced ahead. Archie went to park the car, about 200 yards from the picnic site.

The accounts vary slightly on precisely what happened as Reverend Mitchell exited the car. The overarching points are not in dispute. The picnickers came upon a massive deflated balloon with an attached explosive device. Someone touched the device. It might have been Jay Gifford, who two weeks earlier had won praise from the U.S. Weather Bureau for finding a weather balloon and returning it to the bureau office in Klamath Falls. It might have been Mrs. Mitchell.

Mitchell heard his wife call to him: "Come see this balloon." Although the U.S. government didn't want it known, Mitchell was aware that suspicious balloons were appearing in the Northwestern forests. Instantly, he implored his wife to keep everyone away from the object. It was too late. "There was a terrible explosion," Jumbo Barnhouse would write. Barnhouse would be blown out of his seat. "Twigs flew through the air, pine needles began to fall, dead branches and dust and dead logs went up."[4]

Mrs. Mitchell was still alive when her husband arrived, her clothing on fire. He extinguished the flames with his hands but couldn't save her life. She and the five children were dead.

The horrified Jumbo Barnhouse sped to the ranger station, bailed out of the pickup, bounded up the metal steps, and breathlessly announced to rangers Jack Smith and Spike Armstrong, "There's been an explosion on Gearhart Mountain."

Smith and Armstrong quickly loaded sheets, blankets, and first-aid kits, and raced to the scene. Jumbo headed for Cora Conner's switchboard room.

Smith and Armstrong encountered the stunned Archie Mitchell, his hands stained yellow with picric acid, residue from the flames on his wife's clothing. He directed them to the bodies, which were arranged around a three-foot-wide explosion crater "like spokes on a wheel," in Smith's words. Armstrong deferred to Smith. "Spike said to me, 'Can you check their pulse? I don't think I can do it.'" When I spoke with Jack Smith in 2010, he told me he still had in his possession a shard of shrapnel he extracted from a pine tree at the scene that day.[5]

Select military and Interior Department personnel in Washington knew what Cora Conner and the rest of Oregon, save for a few exceptions, didn't: Mrs. Mitchell and the five children were the first—and, ultimately, the only—World War II civilian casualties on U.S. soil. The Bly balloon was made in Japan. It wasn't the first found in North America; they were sighted as early as November 1944.[6] In fact, archives show it was one of the last, number 240.[7] They were discovered from the Aleutians to Texas to Michigan.

By the time of the Bly tragedy, U.S. intelligence had spliced together the outlines of a diabolical campaign aimed at creating wildfires in the Northwest and panic throughout the nation. On March 10, 1945, one of the balloons had descended close to the building in Hanford, Washington, that was the site of the "Manhattan Project."[8] It briefly knocked out power to the top-secret facility where the atomic bomb, which would cause so much destruction to Japan and end the war, was being created. That six victims knew absolutely nothing about these peculiar Japanese missiles spoke to the success of the U.S. government's secrecy campaign, albeit tragically. Archie Mitchell was among the privileged few to receive a briefing from forest officials; he hadn't even told his wife about it.

As balloon remnants were recovered, they were shipped to military laboratories in California and New Jersey. The analyses established several revealing clues about the construction, operation, and mission of the balloons.

They determined that the canopies consisted of layered sections of paper held together by glue derived from *konnyaku*, a potato-like vegetable.[9] The undercarriages were equipped with an altitude-control device; sandbags; and five bombs, four of which were incendiaries to set off wildfires. To destroy evidence of the mechanism, each undercarriage contained a two-pound demolition block of picric acid, the source of the residue on the hands of Reverend Mitchell.[10]

What the analyses could not explain was how the balloons found their way to North America. After examining one that landed in Montana, naval researchers concluded that "while the balloon is undoubtedly of Japanese manufacture, the evidence thus far available does not permit any definite conclusions as to how and why it arrived."[11] Were the bags being filled with California sand by Japanese frogmen, or by prisoners in U.S. internment camps?

To solve the puzzle of where the balloons were coming from, the military enlisted a group of detectives, far more acquainted with rocks than international terrorism.

Detritus found at some of the crash sites included sacks of sand used to stabilize the balloons in flight and regulate altitude. The quantities of the recovered sand were paltry, but in the hands of investigators from the U.S. Geological Survey's Military Geology Unit, the grains became miniature time capsules that revealed crucial evidence. The grains consisted of microscopic diatoms dating to 1.5 to 5.3 million years ago. They contained tiny skeletons of microscopic organisms that feed on the ocean bottom. Thus, the sand had to come from a beach. But which one? Given what they saw of the composition, the geologists ruled out all North American beaches.

By good fortune, some of the microorganisms matched descriptions found in Japanese prewar geologic papers. The geologists did not detect coral; this was important because that ruled out warmer waters. Further examination of the Japanese scientific papers allowed them to narrow the possibilities to two beach areas north of Tokyo on the eastern shore of Honshu. The launchings would have had to occur near the source of the sand, since sand is so heavy and difficult to transport. So much for the Japanese frogmen or internment camp hypotheses. Just as certainly as the materials were Japanese, the balloons were being assembled in—and launched from—Japan.[12]

The Japanese were exploiting a work of genius. The balloons were riding highways of high-altitude winds about which U.S. intelligence was dangerously ignorant, as was the U.S. scientific community. American meteorologists were distantly behind the Japanese in their understanding of the upper atmosphere. No one in the United States was familiar with the extraordinary accomplishments of Wasaburo Ooishi, an unwitting coconspirator in the tragedy that took the lives of Elsye Mitchell and Cora Conner's friends.

The Esperanto Master

The first link in the chain of events that led to Ooishi's amazing discovery and, ultimately, the Bly bombing, was a boating accident that occurred off the coast of Japan in 1911. A fishing fleet was lost in a devastating storm. A weather-savvy member of Parliament persuaded the government that it needed to invest in research that would improve warnings for storms of such ferocity. To that end, an observatory was established in 1921, at Tateno, northeast of Tokyo, to explore the vast unknown of the upper atmosphere. The logical choice to run it was Wasaburo Ooishi, a Japanese Brahmin who had so excelled at the elite Imperial University that he was given an opportunity to study with Western meteorologists in Germany. He had a particular interest and expertise in the high atmosphere, and had visited upper-air observatories in the United States. He was offered the Tateno post and accepted and executed his mission with enthusiasm. In so doing he demonstrated an almost boundless tolerance for tedium.

He released pilot balloons into the atmosphere, tracking them with a theodolite optical instrument and inferring wind speeds based on their movements. In all, he made 1,228 observations from March 1923 to February 1925. What he documented were typhoon-level winds—more than 150 miles per hour—near the 30,000-foot level, higher than any planes had yet flown. Ooishi had encountered the wind system that would become known as the "jet stream," so named for its focused and dramatic jets of air; the label had nothing to do with aircraft. It was an astonishing finding that he was anxious to share with the rest of the scientific world.[13]

To that end, he published the summary of his pioneering work in Esperanto, the universal language he had mastered and championed, and taught to young people.[14] He was chairman of the Japan Esperanto Society, which promoted a language developed in the 1870s.[15]

Ooishi was as passionate about Esperanto as he was about the winds of the high atmosphere. His granddaughter told me that he was a devotee of "foreign" books, but he believed that Westerners should have no lock on the languages of science. Westerners obviously did not share Ooishi's passion for Esperanto. Outside of Japan, his findings were summarily ignored. In World War II, allied pilots would pay a price for that ignorance.

Ooishi likely was surprised, perhaps appalled, that his research was used for destructive ends. He was no maker of bombs. "He was such a gentle and tender person; I never saw a look of anger on his face," his granddaughter told me in a moving correspondence. (It was translated by *Inquirer* colleague Akira Suwa, one of the most talented photographers I've ever worked with.) Actually, she allowed, she did see him angry at least once, related to violence, violence toward nature, that is.

Yoshitomo Kojoh, who had lived with her grandfather in Tateno after her father died, had not the slightest hint that her grandfather was involved in world-changing research. "I always pictured him as just my grandpa, so I almost could not believe that he worked in the weather science field," she said.

She recalled how Ooishi took the family sightseeing on Tsukuba Mountain to pick *tsukushi* (horsetail shoots) and how "delicious *tsukushi* tasted when my grandmother cooked it with soy sauce."

On the Tateno property, Ooishi cultivated morning glories and chrysanthemums, and was especially fond of his chestnut trees. After he finished his assignment at Tateno, two of those trees were cut down, and when Ooishi came back to visit Tateno after the war and saw what happened to them, he was "very angry," his granddaughter said. "This is the only time I ever saw anger in his face."

She did not learn about his role in discovering the jet stream until decades later. She recalled coming across the book *Balloon Bombs, Documents of the Fugo, a Japanese Weapon*, by Koichi Yoshino, and a 2007 article in a Japanese newspaper. "I knew these were about my grandfather, but I didn't give them much deeper thought than that," she commented. She said her son eventually took her to the weather station in Tateno, where she "met many people who talked about him."[16]

No available evidence suggests that Ooishi was in any way connected to war efforts. His interest was vectored toward advancing solutions to one of Earth's grand puzzles: the workings of the atmosphere. His observation program ended long before the Japanese military conceived the balloon-bombing campaign.

But armed with Ooishi's research, the Japanese were far ahead of the rest of the world in their knowledge of the jet-stream winds. When the Japanese decided in 1942 to go ahead with a plan to launch their primitive intercontinental missiles, they were able to draw on Ooishi's work to map the flow patterns that would become the balloon highways.

Ultimately, those winds that form along boundaries of warm and cold air in the high atmosphere, the winds that ignite and steer storms, are what tethered Japan to Cora Conner and Bly. It was an immense force that neither the Japanese nor some of the best military and scientific minds in the United States understood. Ooishi, himself, did not understand them completely.

What eluded Ooishi and Japan's balloon-bombing strategists was a fundamental characteristic of these winds. They were neither highways nor what the world would come to call a "stream." Rather, the winds behaved more as a powerful river with chaotic and unpredictable eddies and tributaries. Ooishi correctly documented that the winds would be strongest in the cool seasons when the temperature contrasts were most dramatic. What the Japanese didn't consider was that the cool seasons also were the wet seasons in the Northwest, and that factor would greatly depress the fire potential.

What eluded the Allied and Axis strategists was absolutely any acquaintance with Ooishi's work, an ignorance for which they would pay dearly.

World War II remade the world, and it also remade meteorology. Those winds of war were the cosmic forces driving the world's weather. It just took a world war to lure them out of hiding.

"Stupid" Weathermen

A young weatherman named Reid Bryson was stationed on Guam in the fall of 1944, the war at a turning point. The United States was committed to an all-out air assault on Tokyo, dispatching 111 planes to bomb an aircraft factory. Only 24 of them, however, dropped their payloads anywhere near the targets. Pilots were exploring new heights in the atmosphere and encountering what, for them, was an unknown enemy—tornadic-force winds that diverted their bombs. Damage to the ground targets was negligible.

A puzzled and exasperated general approached Bryson and his team of meteorologists for help. Give me a forecast for wind speeds at the crucial 30,000-foot flight level, he ordered. The weathermen were working with paltry data. For them, this was uncharted meteorological territory. From a surface analysis they knew a strong cold front was crossing Japan, which meant strong winds that would increase with altitude. Applying a theoretical formula to calculate wind speeds that no one—to their knowledge—ever had measured, they came up with an improbable answer: 168 knots, or 193 miles per hour.

"Stupid," the general called them. Surely, they meant 68 knots. When our pilots return, the general said, *they* will tell you the real wind speeds.

The results of that mission were "disastrous," Bryson recounted. "The planes couldn't fly upwind because they were practically standing still, . . . sitting ducks for the Japanese antiaircraft fire." A returning pilot told Bryson, "It was strange to see the Japanese coast approaching on the radar, then to see it stand still."

The Bryson forecast wasn't precise; the actual winds were measured at 196 miles per hour. That was close enough to merit an apology from the general.[17]

The pilots had been ambushed by the jet-stream winds that Wasaburo Ooishi had discovered 20 years earlier; the findings were ignored in the West. Now they were being discovered anew by pilots over Japan and also over Europe, the subject of horror stories that airmen would tell later. Those powerful winds bedeviling the bombers intrigued a group of intellectual giants at the University of Chicago who were reinventing the science of meteorology and unraveling the enigmas of the forces behind blizzards and other violent storms in the United States.

"Strahl Strome"

If atmospheric science had a Hall of Fame, the group's impresario, Carl Gustav Rossby, would be a Babe Ruth. "The history of modern meteorology is inescapably paralleled by Rossby's career," *Time* magazine wrote. Rossby came to the U.S. Weather Bureau from his native Sweden. More than a breath of fresh air, he was a howling wind who redefined the state of meteorology. A Swedish colleague predicted, "He would go very far . . . or end in jail."[18]

In Washington, Rossby found an agency that was miles behind the Europeans. Frustrated by life in the bureaucracy, he had a falling-out with the bureau's administration. The final storm blew up when Rossby agreed to work up a forecast for aviation legend Charles Lindbergh, who wanted to fly from Washington to Mexico City. After the bureau chief objected, Rossby quit.

He moved on to start the country's first graduate program in meteorology at MIT and made the fateful decision to run a similar program at the University of Chicago. In his view, Chicago was an ideal location, a crossroads of North American storm tracks. Rossby already had figured out that the world's weather was governed by large-scale upper-air "waves" that control temperature patterns and the movements of storms. Learning of the winds that buffeted the bombers, Rossby and his group turned their attention to the so-called jet stream.

One of Rossby's associates was the brilliant Herbert Riehl, a Jewish escapee from Hitler's Germany in the 1930s.[19] Riehl, classically educated and fluent in English, worked as a stockbroker, screenwriter, and grocer before he obtained U.S. citizenship in 1939. With his citizenship papers, he enlisted in the U.S. Army Air Corp; he was assigned to the meteorology department and sent to New York University, where he graduated first in his class. Riehl later became an instructor at the University of Chicago, where he met Rossby in 1941.

Elmar Reiter, a German meteorologist who quite literally wrote the book on the jet stream—highly recommended for scientists, not beach reading for math-challenged laymen—noted that in the 1920s and 1930s, weather balloons and observations had detected "high-velocity" winds in the upper atmosphere but that they were "locally confined." Once again it was evident that Western researchers had no acquaintance with Ooishi's masterful work, which would have been accessible in Esperanto. The atmospheric ambushes experienced by bomber pilots in the Asian and European theaters were frightening clues that upper-air currents could behave unfathomably. A German colleague, Heinrich Seilkopf, was among those who took note. Reiter said that in 1939, Seilkopf minted the phrase "*strahl strome,*" German for "jet stream," to describe the potent, renegade, focused gales he observed.[20] The image had nothing to do with aircraft—this was well before the jet age—but with high-powered jets of water. That was fitting given that the atmosphere is often described as an "ocean of air," and its behavior often parallels that of the oceans, as Rossby would observe. Seilkopf's label would gain worldwide currency during the next several years, although it was a misnomer.

"I first heard that term when I came to Chicago," Riehl recalled in a personal correspondence. "We were most impressed by the news, that B-29s . . . sent westward to Tokyo became almost stationary." He said that initially the term "jet stream" was used "tentatively but soon was taken seriously, as it indicated not only the speed of a current, but also the narrowness."

Under Rossby's guidance, Riehl and the rest of the University of Chicago meticulously assembled the evidence, analyzed the upper-air charts, and documented that these winds were anything but "locally confined," nor did they constitute a single "stream" around the hemisphere, as researchers had believed. They were planetary, and they were the forces behind the wildest weather on Earth. Riehl was among the first to link the jet stream with blizzards, tornadoes, and cyclones that formed at the surface. For meteorology, this was a profound linkage.[21]

Weather is all about the weight of the air, or air "pressure." Weather reports on popular media these days don't talk much about that most important element of weather observation: barometric pressure, which is a measure of the weight of the 10 miles of air, bonded by gravity, pressing on the earth's surface. It was once popularly understood that changes in barometric pressure indicated changes in the weather. "The barometer of his emotional nature was set for a spell of riot," James Joyce wrote of one of his short-story characters who was preparing to quell his rage with alcohol. Joyce was confident that his audience would grasp the image. Thirty years later, in 1944, that passage was cited in the American novel *Lost Weekend*. Don't expect to see any such references in 21st-century literature.

The "Ls" for "lows" and "Hs" for "highs" still appear on weather maps on TV and online. One Philadelphia TV weather report used to call lows "bad guys" and highs "good guys," but how many people would associate them with differences in the weight of air? To meteorologists, of course, those differences are weighty matters. Heavy air sinks, and the descending currents of highs, flowing clockwise around the centers, discourages clouds and precipitation. Conversely, air travels counterclockwise around centers of lows, and the ascending currents move inward toward the centers. As rising air cools, the invisible water vapor it was holding condenses, falling as rain or snow. When a hurricane-force jet stream wind, five to nine miles high in the atmosphere, interacts with a low at the surface, the result usually is mayhem. The high-speed winds incite the rising air to riot—think of a cork popping off a champagne bottle—transforming an innocuous-looking storm into a blizzard. That's what occurred in March 1888.

In a Spin

The existence of jet-stream winds has to do with the earth's rotation. If the world didn't turn, the lighter, warmer air near the equator, where the sun shines most directly, would simply move linearly toward the poles and vice versa. The east-to-west spin changes everything, and the global circulation is further compounded by the fact that the earth's axis is tilted, and its orbital path around the sun undergoes infinitesimal changes so that the amount of solar radiation reaching any point on the planet is never exactly the same. The fact that the Northern Hemisphere has far more land than the Southern Hemisphere also affects atmospheric circulation. Then there's differing rates of spin.

The earth turns at 1,038 miles per hour at the equator, but halfway to the North Pole at latitude 45—Montreal, where you can buy the world's best bagels—the rate is 734 miles per hour. The rate of rotation becomes slower as you get farther and farther from the equator until it comes to a standstill at the poles. The speed of the winds, however, is the inverse; in general, the farther from the equator, the stronger the winds. The difference between the speed of the air and the speed of the earth's rotation widens as the air spreads from the equator to the North Pole.[22]

"Polar" jet-stream winds, driven by those temperature and pressure contrasts, tend to form about halfway between the equator and the poles; the "subtropical" jet, also important for winter storms, develops just outside the Tropic of Cancer and the Tropic of Capricorn. But these are wild and chaotic three-dimensional wind systems, 4 to 10 miles deep. At their narrow cores, wind speeds can reach 275 miles per hour. They buckle, dip, split, spin off in whirlpools, vanish, and reappear. Is it any wonder they presented such a challenge to atmospheric scientists?

The Agent Provocateur

This was a problem that piqued the interest of Rossby, the force behind Chicago's jet-stream research. And according to Riehl, it was a natural segue for Rossby.

Rossby's interest was rooted in his work in the 1930s on the Gulf Stream, a narrow, high-velocity current embedded in the more ponderously moving waters of the North Atlantic. Rossby at first "met discouragement" in his pursuit of high-velocity cores in the upper atmosphere, but that had to do with the upper limits of his research. His measurements were confined to the 10,000-foot level, the standard at the time. That changed with World War II and the incredible reports of the bomber pilots and the mapping of the upper atmosphere. What Rossby found was that those narrow cores of high-speed winds bore a "startling similarity" to the Gulf Stream, the jet stream's oceanic counterpart.[23]

With any stream, the speed of the current would be highest at the center and weaken the closer it gets to the banks. The jet stream and Gulf Stream obeyed the same principle, and shared other traits, including the fact that while they do behave like them, they are not "streams" per se.

They both owe their existences to the rotation of the earth and share fluid characteristics. They are important weathermakers that are crucial barometers of the future of climate. They are prone to wildly erratic behavior, the Gulf Stream being a "spaghetti diagram of tangled, looping, crossing tracks," in the words of science writer Robert Kunzig.[24] And in very different ways they help explain why New World settlers endured winters that were so much harsher and more volatile than winters in the Old, even in colonies hundreds of miles closer to the subtropics than were the British Isles and the Continent.

The Franklin Current

What Wasaburo Ooishi was to the jet stream, one of America's most famous colonists was to the Gulf Stream in an important sense. Through intuition, observation, and investigation, Benjamin Franklin introduced and explained the Gulf Stream to the world.

He was not the "discoverer"; Columbus knew of the current's existence in some form, although he didn't know it by that name. In 1513, Ponce de Leon, the Spanish conquistador, was the first to provide a written account of his *Carrera de las Indios*, which flowed north along the Florida coast on the west side of the Atlantic. Of his encounter, he wrote, "Although we had a great wind, we could not proceed forward but backward, and it seemed that we were proceeding well, but in the end, it was known that in such wise the current which was more powerful than the wind."[25] On subsequent voyages, 16th-century Spanish sailors would exploit the current to speed up their homeward journeys.

But it would be two and a half centuries before a deeper understanding of the current emerged, and that came from the prodigious mind of Franklin. Contrary to our image of him as the rotund, bespectacled patriarch, Benjamin Franklin was once young and muscular, a 20-year-old sailor. At that young age he already was exhibiting a talent for observation.

On his first transoceanic voyage, in the heart of the North Atlantic, he perceived mild, moist winds on his skin. He noticed that the waters had suddenly developed a deep-blue hue and turned eerily warm. Patches of brown seaweed had risen to the surface. He filed away the observations he made in 1726 for four decades.

In the 1760s, Franklin was the postmaster for the American colonies, stationed in London, and he decided to try to make the mail system more efficient. As a matter of standard procedure, the colonial-bound mail was loaded onto heavy cargo ships that would not leave port until they were full. Franklin added a fleet of smaller, swifter "packet ships" that would leave on their own schedules.

Surely, this change of policy would speed deliveries.

Only, it didn't.

Those slower, balkier cargo ships were completing mail runs to the colonies more than two weeks quicker than the faster and lighter packets. How could this be? One explanation that Franklin heard was that the packet captains were engaged in a work slowdown because they were dissatisfied with their paychecks. It wouldn't be the last time employees adopted such a strategy to extract more money from their employers, but Franklin wasn't so sure about that explanation. He sought a second opinion from one Timothy Folger. Folger was a trusted source; he was Franklin's cousin. He also was informed. He was a Nantucket whaling captain and dealer of sperm-whale oil. He visited London frequently and thus had intimate familiarity with transatlantic crossings.

A decades-old British sailing manual exhorted ship captains to avoid the dangerous shoals of New England. The packets were sailing by the book, taking a southerly route. They were avoiding the shoals, but they were not avoiding the troubled waters of a massive current in the heart of the North Atlantic.

The cargo captains were savvier, opting for routes that avoided the stream. Their packet counterparts, on the other hand, knew nothing about the warm current that was the lifeblood of Folger's whale hunters because it happened to be a lifeblood for the whales. The whales patrolled the borders of the swift current. Plankton flourished at the boundaries of the warmer and cooler water. That drew plankton-eating fish. Plankton-eating fish drew whales. Whales drew whalers.

Folger said that the packet captains were not only uninformed, but also stubborn and arrogant. He said his whalers frequently encountered the packets and advised the captains to stay away from the stream. As Ponce de Leon had learned, even in a favorable wind, the sailing craft would be pushed back by the current. But who were these lowly ragamuffin colonists to tell British captains how to navigate?[26]

Tracing the paths of the whales, Franklin and Folger are credited with drawing the first detailed map of what the whalers called the "Gulf Stream." It was quite an amazing feat and bears a remarkable resemblance to the boundaries evident in those infrared satellite images today. What we know about the Gulf Stream today has a lot to do with generations of whales.

Franklin went beyond charting the geographic boundaries of the stream. A half-century after the tepid waters had first come to his attention, Franklin made another transatlantic trip, this time in the opposite direction. Accompanied by two of his grandchildren, he was sailing to France to seek aid for a project in which he had become enmeshed: the revolt of the American colonies against the British. France and England were not on the friendliest of terms. France was fresh off its defeat by the British in the so-called French and Indian War, and as the aphorism goes, the enemy of my enemy is my friend.

On the journey, Franklin made a major contribution to history—scientific history, that is.

On April 29, 1775, he submerged a thermometer into the deeper-blue waters. He observed an 11-degree jump in sea temperatures, from 60 degrees 14 hours earlier to 71 degrees. He continued taking measurements each day until May 2, four to six times a day, from 7 a.m. to 11 p.m. With each crossing, he added to his dataset.

Of the Gulf Stream, he told a French colleague, "I find that it is always warmer than the sea on each side of it."[27] He encouraged captains to add thermometers to the list of their ships' equipment. Franklin's work marked the beginnings of oceanic observations. The U.S. government would begin taking systematic measurements of the Atlantic in the mid-19th century. Fittingly, it did so at the behest of Alexander D. Bache. Bache was Franklin's great-grandson.

The current that frustrated Ponce de Leon and stoked the interests of Benjamin Franklin is a relentless, meandering mass of warm water that just keeps rolling along—so far, anyway. Some scientists have expressed grave concerns about its future, an anxiety that already has yielded at least one outsized but entertaining movie.

It has been described as a restless snake held by the tail at the Florida Straits. But University of Miami oceanographer Bill Johns, a researcher at the Rosenstiel School, has documented that the source waters extend all the way to the Brazilian coast. Those waters pour through the Caribbean straits and into the Gulf of Mexico through the Yucatán Strait. They swirl into "loop currents," for example, the particularly deep one that ignited Hurricane Katrina. Eventually, the water squeezes through the Florida Straits with almost unimaginable force. Johns likely knows more about the flow of the current than anyone in the world. He has calculated that the volume of water exiting the straits each day would cover a city the size of Philadelphia with a layer of water five miles deep.[28]

The primary path of the Gulf Stream on average is about 60 miles wide, more or less the width of the state of New Jersey. Its speed varies, ranging from two to six miles per hour, resembling an immense, albeit serpentine, mass of water approaching a waterfall. If you have ever tried to paddle a boat against a current of that speed, you can appreciate Ponce de Leon's frustration.

The current churns parallel to the southeast coast, sometimes closer to the shore, sometimes farther away, and makes a sharp right near Cape Hatteras,

North Carolina, an important turn for weather in the East, and continues toward the Iberian Peninsula. Imagine the strength of the Mississippi River and every river that drains into the Atlantic; the Gulf Stream is 300 times stronger. Its power has been likened to that of 2 million nuclear plants.[29] That is why it is one of the planet's most significant heat exchangers, exporting tropical warmth toward the Arctic, a crucial component of the planet's endless effort to maintain a constant temperature. The Pacific Ocean is three times larger than the Atlantic. Yet, the Atlantic transports more heat to the Arctic than the Pacific. In large measure, that's the result of the Gulf Stream.

For climate, the Gulf Stream is a great divide, a major player in explaining why the New World's weather was as different from that of the Old World as the Hudson River was from the Thames.

For all the heat it harnesses, it is ironically one of the world's most important snowmakers. With its atmospheric counterpart, the jet stream, it has been an agent provocateur of nor'easters and blizzards, the ones that so shocked the early European settlers.

Blizzard Fuel

The waters were so invitingly warm that they had enticed a family of dolphins to romp in front of the bow of the government research ship the *Nancy Foster*, named for the marine biologist who worked for the National Oceanic and Atmospheric Administration (NOAA). They splashed in waters velvet-smooth, clear and clean, alternating between a deep-blue and aqua hue.

As a full moon rose, backlighting the horizon, the water transformed the moonlight into a golden mosaic. As the surface responded to a gentle breeze, it generated ripples that scattered the light, reunited it, and scattered it again.

I was fortunate to witness this haunting and magnificent twilight on the western edge of the Gulf Stream off the North Carolina coast while observing NOAA scientists on a research cruise.

Paul Kocin, who has spent his career studying storms, wasn't exactly jealous. Kocin knows that the *Nancy Foster* was prowling an atmospheric minefield.

"It's an extremely dangerous, mystifying area that has a profound effect on weather," said Kocin, coauthor of *Northeastern Winter Storms*, with Louis Uccellini, who now runs the National Weather Service. Said Kocin, who famously deconstructed the Gulf Stream-incited "white hurricane" of 1888. Kocin added, "I try to stay away from it as much as possible."[30]

Throughout the centuries Gulf Stream–related storms have claimed hundreds of ships.

The stream is a celebrated storm-maker, for much the same reason that it has become a focal point of global-warming research: It is a prodigious mover of heat. Only in the last generation have scientists come to appreciate its power, and it continues to surprise them.

They now know that the Gulf Stream has been an agent provocateur in so many of the important winter storms to hit what is today the I-95 corridor. Among them, that historic blizzard of 1888.

Even in this enchanting setting off the North Carolina coast, the Gulf Stream leaves a trail of evidence that hints at its dangerous side. The languid air borne on the current is distinctly tropical and swollen with water vapor. The vapor is palpable to the skin—as Benjamin Franklin observed when he encountered the stream in 1726. It condenses on the cool, white deck railings of the research ship *Nancy Foster*. The Gulf Stream is a prodigious supplier of water vapor, the combustible ingredient that has helped fuel the monster storms.

The Eastern United States is a special place for winter storms, as those colonists would find out. Those storms form at the boundaries of warm and cold air, and the western Atlantic in the vicinity of the meandering Gulf Stream is a particularly fertile breeding ground. Those frigid hurricane-force jet-stream winds interacting with vapor-infused air brewing over the warm Atlantic waters are the perfect combustible ingredients for atmospheric mayhem. In the British Isles and other portions of Western Europe, the Gulf Stream and its tributaries have a tempering effect on those west-to-east jet stream and surface winds. Palm trees grow along the southern Irish coast, which is at a latitude about 700 miles north of Boston's. You will not find palm trees in Boston, Philadelphia, Washington, or Richmond. William Penn was said to marvel at how the Philadelphia sun was stronger and the days lasted longer. Yet, the winter air is more biting than London's, which was 800 miles closer to the North Pole than the Quaker City.

The Gulf Stream's warmth was near and yet so far away for the colonies. It was no friend of the colonial weather. The stream's proximity to the coast helped make their summers more oppressive as a source of humidity and their winters all the more punitive. On several occasions, that Arctic air sliding off the coast near the surface forced the moist parcels over the sea surface to rise; the jet-stream winds in the atmosphere incited them to riot.

The colonists had no way of knowing this, of course. The Gulf Stream's role in storm-making really wasn't well understood until the 1940s, when oceanographer Henry Stommel finally figured it out and was able to explain it.

The Gulf Stream is the western flank of a basin-wide gyre—more or less circular in shape—centered in the North Atlantic. Water flows clockwise around the center, similar to how it moves around high pressure in the atmosphere. Columbus was aware of the gyre. He had learned about it from sailors in the port of Lisbon, with whom he had frequent dealings. On his voyages he had observed exotic tree trunks and so-called horse beans wash up on the shores of the Azores, and seen foreign-appearing dead bodies adrift in a boat near Galway, so strange that the Irish surmised they must have come from Asia.[31] At the time he didn't know of the existence of the Caribbean islands, but he reasoned that only an expansive current system could have been responsible for deliveries over such a distance to the west. The Lisbon sailors exploited south-flowing waters to travel to African coastal regions. Columbus reasoned that the south-flowing current was part of a circular system, based on the appearance of the debris and the corpses. If that were the case, the current would have to make a turn toward the west and eventually bend toward the north and arc back toward Gibraltar.

When he departed on August 3, 1492, and left behind the "gray Azores," to invoke poet Joaquin Miller's phrase, he mined those south-flowing waters. I don't remember how many times I heard in grade school that Columbus's crews lived in mortal fear that the *Nina*, *Pinta*, and *Santa Maria* would fall off the edge of a flat Earth. That was absolute bunk: By the end of the 15th century, that the planet was round was widely accepted. So, what would have driven the sailors to near-mutiny? It was the fear that they would not be able to make the return trip home against the current. Columbus was known to fudge the ship's log so that his men wouldn't know how far they had traveled.

"Sailed this day 19 leagues, and determined to count less than the true number, that the crew might not be dismayed if the voyage should prove long," he wrote in his log for Sunday, September 9.[32] Columbus might have been misguided as to where he was in the world, but he was right about the character of the oceanic current. Ultimately, he was able to return to Spain triumphantly with the powerful assistance from the Gulf Stream, and it is likely that he picked up speed on the way home.

The Gulf Stream current is far more powerful than any other currents in the gyre, and Stommel figured out why. The western flank of the oceanic gyre would be far different from the eastern side. The east-to-west rotation of Earth drives ocean currents toward the west. The south-flowing waters are free to drift leisurely. But the waters to the west live in a far more pressured environment as the planet's endless turning forces them through a narrow channel between the center of the gyre and the continental shelf.

The northward-moving waters to the west come from the warmest places on Earth where the sun is the strongest year-round. Thus, they are laden with warmth, ignition fluid for the frontal-boundary storms that are further uncapped by the high-altitude jet-stream winds. For storm-making power, the only other region in the world comparable to the Gulf Stream's is the Kuroshio current in the far western Pacific, which interacts with cold air moving across Asia. For the Eastern United States, the northern edges of the Gulf Stream and the adjacent area extending several hundred miles toward Newfoundland[33] constitute a primary instigator of explosively developing storms that deepen rapidly. In their 1980 paper, researchers Frederick Sanders and John Gyakum called them "bombs," a label that has stuck.[34]

Gyakum, who teaches at McGill University, in Montreal, has said that Europeans found the term "bomb" offensive because of its association with war. Sanders retorted, what about the term "front"?[35]

The heaviest snow typically falls in the northwest quadrant of storms, often several hundred miles from storm centers. A cyclone that matures off Cape Hatteras and parallels the coast has the capacity to shut down the Mid-Atlantic and Northeast, from Virginia to Maine.

A meteorologist once described winter storms as the "great egg beaters" of the atmosphere,[36] rousting, distributing, and blending radically different combinations of warm and cold air. Winds rotate counterclockwise around storm centers, so the areas to the northwest experience onshore, moisture-laden winds that, in Uccellini's words, "throw back" snow. Often that snow has piled atop land masses inhabited by 50 million people.

While the northwestern Atlantic is a favored region for "bombogenesis," and it is of special significance because the storms that blow up there can have seismic impacts on some of the nation's most densely populated and commercially vital areas, it does not have the monopoly of these meteorological bombs. The Gulf Stream had nothing to do with the tragic "School-children's Blizzard" of January 1888, two months before the deadly sequel in the Northeast.

As Sanders and Gyakum point out, meteorological bombs have detonated on occasion over land, and the jet stream has proved it is perfectly capable of igniting a blizzard without the Gulf Stream's help.

Snow, the Ultimate Math Problem

The pursuit of decoding the jet stream's role and the immense and complicated dynamical forces that generate snowstorms made remarkable gains right after World War II with a deeper understanding of the upper atmosphere. Then came a revolution in forecasting: The prediction of snowstorms would evolve into a giant, and often inelegant and unwieldy, mathematics problem, an improbable science built on probabilities.

As erratically as the atmosphere behaves, it still must obey the laws of physics, its motions governed by cause–effect, action–reaction. It stood to reason that those sequences could be captured by solving a set of equations. The concept of what became known as "numerical weather prediction" was first proposed by legendary Norwegian scientist Vilhelm Bjerknes, whose groundbreaking research documented the role of polar fronts in snowstorms. In a 1904 paper he discusses applying physical laws to forecasting. In his prescient 1922 book *Weather Prediction and Numerical Process*, English scientist Lewis Fry Richardson postulates that the solving of equations of the atmosphere's motions would be the future of forecasting. Richardson envisioned a massive circular hall, a map of the world covering the walls and ceiling, and 64,000 human beings each responsible for weather conditions on a small patch of the world's surface. A "conductor" in the center of the room would coordinate their calculations.[37]

At the time, to get a handle on the next day's weather meteorologists studied sequences of weather maps to predict how the weather would unfold. Say the alignment of "highs" and "lows" on a given day was similar to what it was on March 10, 1888, perhaps that indicated that another Blizzard of '88 was imminent. Richardson saw that type of static analysis—an assumption that what happened will continue to happen—as a dead end for the science. Storms never replicate. As snowflakes, they can be alike, but they are never identical. Meteorology is not astronomy. After the war, Rossby and other adventurers, for instance, Jules G. Charney, who was acquainted with Richardson's work, pursued his vision. Charney had some success at the numerical forecasting of "baroclinic" zones, areas of temperature contrasts where storms form. The western Atlantic is a quintessential baroclinic zone. But the major obstacle to the numerical method was the sheer volume of computations that would be necessary to make this preposterous idea work. Those equations wouldn't fit on a blackboard, not on a couple million of them.[38]

Fortuitously, the postwar interest in the concept of numerical forecasting coincided with the dawn of the computer age and an evermore urgent need for accurate predictions. In 1940, in recognition of the nation's shifting economic realities and the need to direct more attention toward air-travel and automotive interests, the U.S. Weather Bureau was transferred from the Department of Agriculture to the Department of Commerce.

One of the prime promoters of the concept of the numerical weather forecast was computer pioneer John von Neumann, a titan among 20th-century mathematicians, who saw it as both a splendidly challenging and practical problem. Von Neumann, who also was familiar with Richardson's work, in 1946 proposed using his computer facilities at Princeton University for the computations. The administration of President Harry S. Truman bought into the idea. Along with Rossby, Charney, and associates, von Neumann was able to persuade the government to fund a computer-modeling project and establish the Numerical Prediction Unit.[39] By the mid-20th century, the Weather Bureau, led by Francis W. Reichelfelder, one of the first American meteorologists to embrace Bjerknes's theories, was at the vanguard of a forecasting revolution. It couldn't happen soon enough, for in 1953, the bureau would be embarrassed once again, this time from its own backyard all the way to the New York state Snow Belt.

On November 6, 1953, the official government forecast for Philadelphia and vicinity called for mostly cloudy skies with a high of about 50. That day and into the next, more than 13 inches of snow fell on Philadelphia, at least according to one very credible weather observer who would one day figure prominently into the nation's weather industry. We will talk more about him in a subsequent chapter.

For Philadelphia, it would be not only the heaviest snowfall for early November, but also the biggest ever before mid-December. The stripe of heavy snow cut through Pennsylvania to upstate New York. In Buffalo, the official total was 15.3 inches, with 27 inches measured just to the north of the Keystone State's capital, Harrisburg.[40] In the nation's capital, where Reichelfelder worked, 6.2 utterly surprising inches were recorded officially, and close to four inches fell in New York City's Central Park.

The storm originated in the Gulf of Mexico, moved off the Florida coast, and made a hard turn northward and off the coast, traveling directly over the Gulf Stream. Unlike other nor'easters, this one didn't sail farther out to sea and dissipate over the cold waters of the far North Atlantic. Rather, it made an unusual turn inland near the New York harbor, almost like Hurricane Sandy's in 2013.[41] It generated reported hurricane-force gusts of 98 miles per hour on Block Island, Rhode Island, and 78 miles per hour on Long Island. The storm plowed diagonally across New York state, spinning out heavy snows to the north and west of the center, and pouring rain in New England, which was on the warm side of the storm's circulation.

A weather service official allowed that the storm was unique. That was true. Yet, while its path was extraordinary, that would hardly explain a forecast of "mostly cloudy with a high of 50" in Philadelphia. The atmospheric ambush also set off an internal storm. Reichelfelder ordered an in-house investigation. Not surprisingly, the forecast maps had been dangerously off-target, and forecasters had failed to keep up with the rapidly changing, volatile conditions.

The world of forecasting, itself, was rapidly changing. Seven months after that storm, Reichelfelder's agency and the U.S. military undertook an initiative that would synthesize the visions of Richardson, Ooishi, Bjerknes, Rossby, von Neumann, Charney, and others. In July 1954, deeming that the state of computer modeling and power were sufficiently advanced, the U.S. Weather Bureau, U.S. Air Force, and U.S. Navy established the Joint Numerical Weather Prediction Unit. The era of computer-generated forecasting was underway. The future of weather forecasting was in the solutions of equations.

The era of weather surprises was over.

Right?

CHAPTER SIX

Of Bombs and "Flakegate"

When Forecasts Fail

By the end of the 20th century, stunning advances in computer and atmospheric sciences had wrought what had once been the inconceivable. On January 18, 2000, John J. Kelly, head of the National Weather Service, boldly proclaimed that the addition of a new supercomputer—a 786 processor IBM System Parallel (SP) known as the IBM SP, put us "closer to reaching our goal of becoming America's no-surprise weather service," invoking Holiday Inn's once-famous advertising slogan.[1]

The National Weather Service offered one of those numbers that, in reality, 99 percent of its "users"—the likes of you and me—would find about as meaningful as those salt-tonnage figures highway departments throw out there. When we hear they are spreading 10,000 tons of salt, I guess we are supposed to feel safer, or at least reassured that our taxpayer dollars aren't being spent on junkets and cocktail parties.

When Kelly said that the IBM SP was processing 690 billion "instructions" a second and was five times faster than its predecessor, it certainly sounded impressive. The clear implication was that what humans might miss, the machines wouldn't. This wasn't 1953 anymore. Had the government at long last discovered the cyber Rosetta Stone?

The answer would come a week later.

Surprise Party

On January 24, government and commercial meteorologists predicted that a low-pressure area in the Southeast would go on its plodding way into the

Atlantic and stay far enough offshore to pursue a forgettable career as a harmless "fish storm." But that evening the system underwent a rapid intensification off the Carolina coast, and on the morning of January 25, tens of millions of Easterners woke up to a ferocious, life-interrupting snowstorm. Raleigh, North Carolina, and the Golden Triangle were shut down under record 20-plus inches; totals of 10 and 15 inches were common all the way to New Hampshire. So much for the "no surprise."

The January 2000 fiasco was the result of a computer-programming glitch, according to Gary Szatkowski, who was the meteorologist in charge of the National Weather Service Office in Mount Holly. That office is responsible for forecasts for Delaware, eastern Pennsylvania, part of Maryland, and two-thirds of New Jersey, where about 20 million people live. Just about his entire domain fell within the humiliation zone. Szatkowski was never one to shy away from going public to take responsibility for a busted forecast, something that would one day get him in hot water with the big boss, Louis Uccellini, who might well be the world's number-one authority on winter storms. When forecasts went awry, Szatkowski always was quick to pick up the phone and answer for it.

In this case, Szatkowski said, upper-air winds in the Southeast intensified rapidly on January 24, a dead giveaway that the storm was gaining strength. Unfortunately, they were so much stronger than earlier readings that computers rejected them; as a safety measure they were programmed to exclude outlier data.[2] One person who correctly predicted that snow was coming to Philadelphia was TV meteorologist John Bolaris, who a year later would end up with a target on his back due to a busted forecast; we will discuss that later. During his segment on the late news on the night of January 24, he looked at the radar trends and told his audience that computer guidance be damned, snow was coming because nothing was going to stop it. The heads-up was too late for those who already had gone to bed, and presumably too late to inform the hundreds of thousands of schoolchildren who would wake up to one of nature's greatest gifts—a day off from school—a sentiment not necessarily shared by all parents.

Meteorologist Elliot Abrams, a national radio-voice legend at Accu-Weather, which along with The Weather Channel is an Amazon of the commercial weather business, held that the computer absorbed far too much of the blame: The National Weather Service had raised expectations with Kelly's public announcement. In fairness, the National Weather Service's January 18 news release said it wouldn't reach its peak power until September, when its capacity would have more than tripled to 2.5 trillion "instructions."[3]

It is safe to say that the term "no surprise" got far more attention than the fact that the system would need several months of ripening.

At least the computer would be ready for the following winter, and the new millennium would open onto a new scientific epoch for the National Weather Service—the successor of an agency founded in the 19th century, when weather forecasts were the products of papers and pencils, and hand-drawn fronts and isobars—and the future of predicting what the atmosphere would do next.

That first winter of the 21st century did turn out to be quite an eventful one for the computers and forecasters.

The Winter of Discontent

Louis Uccellini had seen the signs as early as Wednesday, the last day of February 2001. At the time he was head of the National Centers for Environmental Prediction, which operates out of a nondescript office building in Silver Spring, Maryland. He was responsible for all winter-storm forecasting and the government's centers of mayhem: the Storm Prediction Center and the National Hurricane Center. But winter storms are his forte. Very possibly no one on Earth knows more about them or talks about them with more enthusiasm than Louis Uccellini.

On February 28, it was clear that the global forecast models were onto something ominous, and two days later Uccellini's staff boldly issued a statement warning of a storm of "historic proportions." It was to start on the following Sunday, and it threatened to paralyze the I-95 population corridor with a 48- to 72-hour siege of one to three feet of snow. The government warned of the potential for "extreme" conditions, with high winds and highway-closing drifting snows.

Fortuitously, it so happened that Uccellini was scheduled to speak to a group of parents and students at West Chester University on Wednesday, March 7. I made it a point to be there. Uccellini looked forward to basking in this triumph of science. That's assuming that nature and snow-fighting crews would allow him to get there for his talk. He would not be wanting for material. West Chester is in Chester County, about 25 miles west of Philadelphia, and 80 miles from the Atlantic Ocean, and the wealthiest county in Pennsylvania and one of the most rapidly developing. It looked to be ideally situated for this particular assault; the county has endured some profound late-season snows.

A snow and rain mix along the coastal plain might be all snow in Chester County. The county is situated along the so-called fall line, the subtly rising areas where the coastal plain yields to the Piedmont (foot of the mountains) Plateau from Georgia to Maine. Temperatures decrease a degree Fahrenheit or so for every 300 feet of elevation, and that can be a crucial difference when readings are close to freezing, as they often are in March storms. Warm air from the ocean borne on easterly winds has a harder time getting to the fall line, and the slight elevation can give the air a convective lift. In a March 1958 storm, a Snow Belt–worthy 52 inches was measured in the northwestern end of the county.

Not only the timing, but also the venue for his talk was fortuitous. This would be the counterpoint to the December 2000 fiasco when Chester County residents had a surreal winter experience. Yes, no doubt the audience would want to know what happened, and Uccellini would come prepared to talk about that one also.

His first order of business would be to explain what happened in the December case. On the day after Christmas, the models run by the United States, Canada, and the United Kingdom agreed that a fast-moving "Alberta Clipper" from the Pacific would roar out of southwestern Canada and interact with the Atlantic to ignite a potent cyclone off the East Coast. With each day, the forecasts became more certain and more ominous.

On Friday, December 29, TV meteorologist Kathy Orr in Philadelphia warned the audience that on Saturday the storm "bombs out," over the ocean. The forecasts called for as many as 15 inches of snow throughout the viewing area, which included West Chester. "Bombs out . . . sounds serious," the news anchorman declared soberly.[4] Heavy snow was expected from Washington to New England and likely would create nightmares for anyone who had ambitions to travel to celebrate the beginning of the third millennium. It became known as the "Millennium Storm." (Purists, including my persnickety former managing editor, affirm that the millennium actually begins in the '01 year, not '00; people count from 1 to 10, not 0 to 9.) Philadelphia was in the computer model bull's-eye. One sure sign of that was the fact that The Weather Channel sent its most visible personality, Jim Cantore, to the city for a Friday night stand-up in front of the steps of the Philadelphia Museum of Art, made famous by movie pugilist Rocky Balboa. Cantore allowed how the "smell" of snow was unmistakable in the Philadelphia air. Philadelphians no doubt were forever grateful that he avoided any temptation to say the storm would deliver a "knockout blow."

The snow did arrive, but not everywhere. The western half of Chester County received almost nothing; an atmospheric dry wall shut off snows along a sharp boundary not far to the west of the I-95 corridor. One Chester County acquaintance reported that she saw snow falling in back of her house and nothing out the front. The dry wall also shut out the Baltimore–Washington area, making a mockery of the forecasts, and, Uccellini said, it prompted a derisive inquiry at his house. "My son said to me, 'What is it you do for a living again?'"

Uccellini offered no excuses—that day or any others, in my experience. Throughout the years he has held to the position that his meteorologists do their best with the tools they have and that one could not ask for more. He did offer an intriguing explanation for the forecast flub. December had been especially cold in the East, and those frigid air masses plowing southeast from Canada had chilled the nearby ocean waters west of the Gulf Stream. He displayed a slide showing the sea-surface temperatures. The waters off the Carolina coast were significantly chillier than the computer models were led to believe.

Worldwide observations at a given time provide an "initial" state of the atmosphere, which is compared with conditions six hours earlier to calculate how it might change in six-hour intervals on out in time. In this instance, the models relied on faulty ocean-temperature data that was based on estimates, not real-time measurements. In turn, the computers overestimated the storm's potential to blow up off the Southeast coast. The cyclone did not mature until it encountered the more favorable dynamics farther to the north than was forecast, having the effect of yanking the snow shield to the east. It also was a bust in Boston, where the forecast the morning of December 30 called for 5 to 10 inches. The counterclockwise circulation around the storm center imported more warm air across eastern Massachusetts than expected—Boston is at a longitude that is about 180 miles east of New York City—and snow accumulations were negligible.

Regardless of how it was viewed in Boston, Washington, or Chester County, the Millennium Storm forecast overall was a success. Philadelphia, itself, reported nine inches, and New York, a foot.

Consider that computer models generate forecasts for every level of the atmosphere for the entire surface of the planet, all 195 million–plus square miles of it. My favorite image likens forecasting the weather to predicting what will happen to water in a full bucket being carried by someone who has downed six shots of tequila.

The atmosphere is a three-dimensional gas that behaves like an unevenly heated fluid, attached rather loosely to a sphere spinning at 1,000 miles per hour hurtling through space at 67,000 miles per hour.

To miss the location of a center of low pressure by a mere 50 or 100 miles two days out—even one day—should be considered an amazing accomplishment in the court of common sense; however, that court is never in session along the densely populated corridor from Washington to Boston.

Easterners might view themselves as sophisticates, but they tend to be myopic when it comes to distances and weather. A 300-mile drive through Wyoming and Utah might pass through towns with a total population of 1 million; on a drive from Washington to Boston, make that 50 million. When small errors result in massive differences in impacts, those inhabitants are unlikely to celebrate the marvels of computer models when small errors result in massive differences in impacts. People want to know when and how much where they live, where they are, and where they're going. Said Greg Postel, a veteran storm specialist at The Weather Channel and a native Philadelphian, "It is indeed the case that the expectation has outrun the science."[5]

While Easterners' demands of meteorology might be unrealistic, their concerns do have bases in experience. The East Coast has endured storms of extraordinary magnitude, and the impacts actually are magnified by the fact that they aren't frequent. The street departments of the coastal plain cities aren't as well armed or thoroughly seasoned as those of the upstate New York Snow Belt cities of Syracuse, Buffalo, and Rochester. Thus, coastal storms can be particularly hazardous to public safety and the economy.

Similarly, forecast "busts"—even grander-scheme-of-things successes that are viewed as busts in areas that get shortchanged on snow—can be disruptive, not to mention publicly embarrassing for meteorologists. A mere forecast for a major storm sets of a chain of responses—cancellations, school and business closings, airline diversions, public-transit delays and reschedulings, supermarket panic—all of which might be viewed as having been absolutely unnecessary by the snow-deprived. They are apt to blame the messengers. These tight geographic margins of error are what the government and commercial forecasters are up against with every East Coast storm. Nowhere else in the world could they come so close yet be so wrong.

But in the first winter of the 21st century, they were armed with the government's new supercomputer, whose unprecedented power had ripened. Chester County didn't get its snow on December 30, 2000, but 40 million other people did, and they knew it was coming days in advance.

Two months later, Uccellini and his forecasters were poised for another dramatic success, this time one that would include Chester County, and Uccellini would be there to declare victory.

The Conquered Hero

The weather did not stop Uccellini from making his way to West Chester for that speech on March 7, 2001. Thus, he did get that opportunity to discuss his passion, although he did not get the opportunity to luxuriate in his triumph. "I thought I would come here as the conquering hero," he told the audience. The storm was another bust in Chester County, and this time for the entire Washington to New York corridor. It was a bust of "historic proportions." The *New York Post* tauntingly declared, "March Lion a NY Pussycat." The paper noted that travel was indeed a disaster—not because of the weather, but because of the forecast.[6] Airlines had moved swiftly to cancel flights preemptively. Manhattan ended up with less than four very manageable inches, and rather than one to two feet, Philadelphia experienced barely an inch of quasi-frozen crud. Uccellini pointed out that this was a mega-storm in upstate New York—30 inches was measured in Troy, 40 inches in northern New England. On a planetary scale, the forecast didn't miss by much. But for Washington, Baltimore, Philadelphia, and New York City, it wasn't a "meteorological bomb" but a bomb of another type.

The Price of Success

While both the December and March storms presented very different forecast challenges, ironically, they both highlighted the enormous gains in forecasting technology that had raised the potential for accuracy—and frustration.

On March 1, 2001, the day after Uccellini and his forecasters had seen the signs of impending trouble foretold in the computer guidance, the *Inquirer*'s top editor asked me if I was planning to write about the snowstorm that was coming Sunday. I was nonplussed. I reminded him tactfully that it was only Thursday and that computer models have been known to hallucinate four days before the scheduled arrival of something that only they were seeing. Nothing is more efficient at removing snow than a computer model. The boss had seen a "crawl" on a local TV station about a potentially crippling snowfall and promised viewers that meteorologist John Bolaris would tell them all about it on the eleven o'clock news. Bolaris became the lightning rod for the failed forecast, and I would argue unfairly so. He recalled in a published interview that he had nothing to do with that promo. The late-news show on his station, Channel 10, had just secured the number-one rating. The Philadelphia market was ferociously competitive, and nothing could rival an oversize snow threat for luring an audience. The drumbeats got only louder the remainder of the week, and Bolaris was hardly the only one beating the drums.

On Friday afternoon, the forecasters in Uccellini's shop, the normally staid National Weather Service, posted a discussion warning of the potential for a storm of "historic proportions." Bolaris, the market's most well-known and animated television meteorologist, was on top of his game. He invited me to come to the studio to watch him and his colleagues grapple with the atmosphere during the weekend. Before I left work Friday, I did agree to write a forecast story that would appear in the Sunday paper. I am ever wary of forecast stories and always try to write something that will stand up no matter what does or doesn't happen. The article was built around the forecast process and sought to explain why almost every forecast contains a 100 percent chance of at least some uncertainty. It was cautiously worded, advising readers that the storm wasn't a done deal. That was the version that appeared in the "bulldog" edition distributed to newsstands and convenience stores early Saturday afternoon. To my chagrin, the top of the article was rewritten later Saturday for the home-delivery editions, our primary circulation runs. It emphasized the coming of the storm, warning of possible "blizzard" conditions, under a headline that screamed, "Massive snowstorm heads toward region; up to two feet predicted."[7] The story still contained several paragraphs of discussion about what could go wrong with the forecast, but readers had to go to the jump page to find it.

The snow did begin more or less on schedule early Sunday afternoon, albeit tentatively. A few hours later, it stopped. Forecasters said don't be fooled, it's still coming, it's just delayed. Our front-page headline on Monday read, "Storm holds off—for a day."[8] A storm did come—a storm of public outrage from the nation's capital, to Independence Hall, to Wall Street.

How outraged? Bolaris said he received death threats.[9] Bolaris was an unabashed storm geek. Most meteorologists are. On TV they attempt to project somber tones about snow threats and promise they'll be with the viewers to provide an electronic counseling service throughout the coming trauma. If a snowstorm whiffs, they're likely to say something on the order of, "We lucked out." But they aren't fooling anyone. Most of them are rooting for that extreme event that will lead to an undeclared public holiday. Bolaris probably did a worse job of concealing his enthusiasm than most, but in March 2001, he became a target, in large measure because of his visibility, the number-one weatherman on the number-one station, his celebrity only heightened by his very public relationship with the attractive young anchorwoman. What he reported on that Wednesday night newscast simply was based on computer guidance available to every forecaster in the world.

For that Sunday story I had spoken with his staid colleague, Glenn Schwartz, a highly regarded meteorologist who was a product of the Penn State weather factory. On the air, Schwartz was the mirror opposite of Bolaris, under orders from the brass. He said he was told that for marketing purposes, he was to be the "anti-Bolaris."[10] He wore a bow tie and glasses, and included healthy heapings of science in his forecasts. Schwartz defended his colleague and all the meteorologists who drew attention to the computer guidance that suggested a coming mega-storm, at the same time recognizing that those miraculous machines have a talent for error, as well as prescience.

"In a way you're setting yourself up for criticism," he said. "At the same time, why should we not share what the science is telling us?"[11]

What's Missing

So, what was the science not telling us? What were the computers missing?

Again, Uccellini did not apologize to the West Chester audience for what to them was the second forecasting fiasco of the winter. He emphasized that the computer guidance wasn't all that far off about where the storm ultimately would mature. That represented quite a technological achievement. It is safe to say that about 25 million people in the Washington to New York corridor were not impressed. Their snow was sabotaged by the seed system over the Pacific Ocean. It took its time traveling across the country and interacted in unforeseen ways with the eventual Atlantic coast storm in its early stages of development. Uccellini told the audience that the interaction, in turn, had the effect of slowing down the system and allowing snow-mitigating warm air to intrude along the coastal plain.

The computer models had whiffed on the impacts of the Pacific system. The Pacific covers about one-third of the earth's surface. It has a complicated relationship with the overlying air and the jet-stream winds that ultimately affect weather in North America. Capturing upper-air conditions over the North Pacific is absolutely crucial to forecasts in the United States. Unfortunately, almost nothing in the atmosphere was more elusive. The atmosphere over the North Pacific, said Uccellini, existed in a "data void." And therein was the number-one problem with that forecast and all others.

One of the paradoxes of the forecasting universe is the simple fact that computing progress has far outpaced progress in the fundamental exercise of observation, and personally I don't think the meteorological community has done enough to make the public understand that weakness and just what a complication it is for any forecast. Thoroughly knowing the present state of the atmosphere is crucial to predicting how it is likely to change throughout time. For a variety of reasons, including the obvious fact that 70 percent of the planet's surface is covered by water, capturing that elusive "initial" condition remains a challenge, and meteorologists readily point to that weakness as the Achilles' heel of all forecasts, both on a short-term and long-term basis.

To reliably predict what the atmosphere is going to do it is essential to know what the atmosphere is doing now.

When Data Holes Are Canyons

The pilot was on the prowl for an atmospheric spell of riot. He found it, a jet-stream gale of 161 miles per hour, roughly the peak wind of Hurricane Andrew when it landed on the Florida shore in 1992. That reading was relayed from a "dropsonde" bomb-like device ejected from the *Gulfstream IV*, a jet commissioned by the National Oceanic and Atmospheric Administration (NOAA) to probe the secrets of the North Pacific data void. I was aboard a flight during the winter of the record snows in the East in 2010. The bomb, 18 inches long and 2.5 inches in diameter, acted like a weather balloon in reverse. As are the balloons that are launched each day from hundreds of ground stations throughout the world, the probe was equipped with an instrument pack that relayed vital information as it descended 45,000 feet. On the descent, a parachute opens and the bomb falls at a rate of two miles per hour, taking 13 minutes to splash down in the roiling waters of the North Pacific. Weather balloons profile the atmosphere from the ground up and the bombs from the top down.

Ascending balloons relay vital, one-of-a-kind data that inform computer models throughout the world. It is an amazing international network, but the coverage is incomplete. Lesser-developed countries tend to have more vital priorities than gathering weather data for computers. It helps to have a stable government. (In Afghanistan, in the mid-1990s, the Taliban shut down the national weather station, which had been well equipped and staffed during the Soviet occupation, because weather forecasting was viewed as "sorcery."[12]) While the atmosphere above land masses is incompletely sampled, the oceans, which cover 70 percent of the planet's surface, present far greater challenges.

The upshot is that because of the world's observation holes, computer-model runs are handicapped before they get started. Without getting too far into the thickets of geek-dom, computer models are dependent on capturing that all-important "initial condition," the state of the atmosphere at a given moment. Models compare that "snapshot" with the state of the atmosphere six hours prior to predict how it will change in the next 6, 12, 18, and 24 hours, and so on, on out in time. But that initial condition is always portrayed imprecisely; it's not as though someone says "smile" and the atmosphere stands still for a portrait. It is a colossal moving target, layer upon layer of air masses with wills of their own, that 10-mile-deep three-dimensional sloshing fluid, attached to a sphere spinning 1,000 miles per hour and hurtling through space at 7,500 miles per hour. It would be impossible for any computer to somehow foretell the progress of every macroprocess and microprocess—for example, evaporation and cloud-building—and how they might affect sensible weather.

Edgar Lorenz, a professor at MIT who died in 2008, is credited with being the first to caution that the seemingly insignificant disturbances in the atmosphere can have tremendous and unpredictable consequences. His hypothesis appeared in a 1972 paper entitled "Predictability: Does the Flap of a Butterfly's Wings in Brazil Set Off a Tornado in Texas?" He became the father of "chaos theory."[13]

If the unseen, unmeasured, and unresolvable are the great enemies of reliable forecasting, what's to be done about them? Computers end up obeying a variant of jazz trumpeter Miles Davis's advice to "play what's not there." Meteorologists rely on "data assimilation" and "parameterization" to compensate for the weaknesses in data and resolution. They rely on "ensemble" forecasting, tweaking models and running them with slight variations in initial conditions and in how they resolve various physical processes. The aim is to arrive at forecasts with the highest likelihood of verifying, ultimately presented in a forecast language that the public can understand.

Often the models perform quite well—some better than others; the models run by the European Community are considered the platinum standard, but they, too, have had their stumbles recently. Different models use different calculations to account for those gaps, which is why one model's historic snowstorm at day 5 is another model's "partly cloudy." Satellite reconnaissance is invaluable, but clouds do get in the way, and the data from space does not have the resolution of on-site measurements. A forecast for a given area is reliant on getting it right through all layers of the atmosphere; if it's wrong about what's happening 45,000 feet above the surface of the Pacific, that increases the chances that it's going to be off the mark three days later in New York City.

In short, there is nothing like the real thing, and that is why the *Gulfstream IV*—an apt named given the relationship between the jet stream and its storm-making counterpart in the Atlantic—is hunting the skies for storm-inciting gales. The crew had left the hotel at daybreak, stopping at a Japanese American delicatessen for sandwiches for what would be a daylong flight. After a briefing at which the crew was advised it might encounter winds of 120 miles per hour, the plane took off from a remote, highly secured area at Honolulu Airport. The jet would trace a snowshoe-shaped pattern and drop 16 probes. The crew was stalking a monstrous storm affecting coastal areas from Alaska to Central California, battering southwestern Canada with hurricane-force winds. About 1,600 miles northeast of Honolulu a probe relayed a wind reading of 138 miles per hour, a significant disparity from the computer-model forecast of 120, and those winds wouldn't be the worst of it; the cyclone was more ferocious than anticipated. The gales were howling on the eastern side of the upper-level system. In about 72 hours, estimated on-board meteorologist Jessica Williams, the same disturbance would race across the country and set off a "heavy precipitation" event in her hometown, Willow Grove, about 20 miles north of Philadelphia, which already had received a record amount of snow that winter. She turned out to be right; perhaps that forecast prospered from the data mined by the *Gulfstream*.

What about the Atlantic?

On the other side of the continent, the Atlantic presents an entirely different set of challenges to the collection of that data that ultimately is ingested by the high-powered computers that produce forecast models. The Millennium Storm of December 30, 2000, demonstrated the impact that Atlantic sea-surface temperatures can have on storm development and how failing to portray them accurately can skew details that make all the difference in the outcomes for tens of millions of people. It is the complex interactions of cold air pouring off the continent, high-speed westerly winds in the upper atmosphere, and the warmer air over the ocean that ignites those East Coast blizzards. And for mega-winter-storm development, no part of the Atlantic is more important than the Gulf Stream.

"That's why they like us out here," says Captain Jamie Velarque. He is watching the weather closely from the bridge's picture windows, which look out over the actual Gulf Stream, not to be confused with the Pacific-prowling jet. (It might be symmetrically appropriate if his ship were called the *Jetstream*, but this is the *Nancy Foster*.)

In addition to ferrying scientists on research cruises, it had another important mission: to send back live data from the stream, another area where "ground truth" observations are sparse. On this particular night, for example, satellite readings indicate a water temperature in the upper 70s; the ship's thermometer is showing 84 degrees all the way down to 300 feet.

Unfortunately, even if the measurements at the sea surface and throughout the atmosphere over the oceans and all the land masses were precise and resolute enough to capture the initial conditions perfectly, and supercomputers could process all the information and resolve all those elusive micro-processes, no one reading this is likely to see the day when forecasts will be foolproof. The atmosphere is what scientists call a nonlinear chaotic system, which is probably putting it mildly. Edgar Lorenz's butterflies are always flapping their wings somewhere in the world. Errors large and small will always threaten to contaminate the purest datasets, and out in time those errors will only grow. And again, this is something that atmospheric scientists need to communicate to the public.

In light of the obstacles, meteorologists and the scientific community have executed some remarkable successes in the last 50 years in which numerical forecasting has become standard operating practice. And ironically, what made the failures of the first winter of the millennium so baffling was the fact that the National Weather Service had scored a sequence of forecast triumphs, foreseeing the development of major snowstorms days in advance and coming close to estimating the areas of impacts. In that context, Kelly's "no surprise" comment wasn't so surprising.

Super Storms and Super Success

In February 1983, government meteorologists were well out in front of private forecasters in predicting the blizzard that would heap what was then a record 21.3 inches on Philadelphia and 5 inches of snow in just one hour in New York City, trapping motorists inside the Lincoln Tunnel for eight hours. Andy Gregorio, then the meteorologist at the venerable Franklin Institute in Philadelphia, who had called for a more conservative 6 to 10 inches, at the insistence of his superiors, conceded that the performance of the government meteorologists was outstanding. "My hat's off to them," he said.[14]

The government evidently had learned some things following the pratfall of the Presidents' Day Storm of 1979. A harmless-looking storm mutated into a 20th-century White Hurricane before the very eyes of National Weather Service headquarters. The nation's capital was entombed under almost two feet of surprise snow, as was Atlantic City, where the era of casino gambling had begun eight months earlier. In Philadelphia, expected to see nary a flake, almost 15 inches was measured.

"It was supposed to be this little system," recalled Paul Kocin. "People went to sleep, and when they woke up they wondered, 'Why can't I open the door?'"[15] The recipe was a familiar one: frigid air sliding off the continent over the warmer sea surface, getting a powerful lift from potent jet-stream winds. What gave the storm its extraordinary ferocity, however, was its interaction with the Gulf Stream, recalled Chet Henricksen, a retired National Weather Service meteorologist assigned to headquarters at the time. The cyclone became so intense that in satellite imagery a hurricane-like "eye" was clearly visible. He had the misfortune of being the lead forecaster who lived closest to the office, and he was the one who got the call after midnight with orders to head into the office because a storm was blowing up off the coast.

The weather service, and everyone else, had to play catch-up. Henricksen was interviewed the next day by a *Washington Post* reporter who relayed that a private-sector forecaster had said the government had blown the forecast and that his outfit had outperformed the National Weather Service. "He's full of it," Henricksen told him, and that comment found its way into the headline.[16]

Every winter, storm forecasts are bound to misfire. Yet, throughout the 1980s and 1990s, neither the National Weather Service nor commercial meteorologists suffered an ignominy that ranked with the Presidents' Day miss. In fact, to its 1983 resume, they added two spectacular hits in the 1990s that affirmed the tremendous improvements in international computer modeling. Computers foresaw a true "Storm of the Century," an all-out blizzard, with devastating impacts from Florida to Maine, from the Atlantic Coast to the Ohio Valley, a storm affecting almost half the nation's population, unprecedented in areal coverage. The guidance suggested the potential for hurricane-force winds and two- to four-foot snows. These were shocking possibilities, given the generally quiet nor'easter seasons of preceding winters. The "Perfect Storm" of 1991 was a hybrid of tropical and mid-latitude cyclones. It clocked New England, erased sand from beaches, and famously roiled the North Atlantic, but it had lesser impacts on coastal areas to the south. An impressive one in December 1992 had ripped sand off beaches and pounded the I-95 cities with gale-force winds, but it was a coastal event that occurred while ocean temperatures still were shedding the warmth of summer and fall, and was not a prodigious snowmaker in the most populated areas.

They did not come close to matching the potential of the Blizzard of 1993.

The tom-toms began sounding loudly at the beginning of the workweek on Monday, March 8, 1993. In its 2:30 p.m. discussion on Monday, March 8, 1993, the government's National Meteorological Center mentioned the threat of "heavy snowfall" during the weekend from the southern Appalachians all the way through eastern metropolitan areas. The center came to that conclusion based on model outputs from the European Community, the United Kingdom, and its own model for the mid-range outlooks forecast, known as the MRF. On Tuesday, forecasters predicted that on Saturday morning, the center of the storm would be over southwestern Georgia, and that was mighty close to what actually happened. By Wednesday, meteorologists were all but certain that something very big was brewing. The center's 2:30 p.m. discussion stated, "There is little doubt there will be very heavy snowfall all along the Appalachians from Georgia to New England."[17]

Perhaps no winter-storm forecast in history had triggered such a massive response, as emergency managers made their preparations and the alarmed East Coast populace sacked supermarkets for life-sustaining supplies—milk, eggs, bread—like Visigoths attacking Rome. Nothing quite matches an outbreak of East Coast prestorm panic.

What the public didn't know was that the models predictably were squabbling, changing their minds about the path of the storm and its intensity, and the other systems that could interact with it and alter its destiny. By now, from hard experience, the models' interpreters expected such vacillations, and to the credit of government and commercial meteorologists, including those at the big shops, The Weather Channel and AccuWeather, they let the models do their bickering over the details, allowed them to luxuriate in their run-to-run whims, but never vacillated on the concept that this would be a storm of legendary magnitude.

One early indication that the meteorologists had made the right call came as the storm crashed into the Florida Gulf Coast, raising water levels to record heights.[18] Its winds battered the National Hurricane Center building in Coral Cables, ripping instruments from the roof; the government later decided it would be wise for the hurricane center to change venues.

The captain of the *Oleander*, a cargo ship that also provided valuable real-time measurements from the Gulf Stream, recalled that 55-foot waves forced it to cancel its trip to Bermuda. The U.S. Coast Guard rescued more than 100 people from stranded ships. Heavy snows fell from Alabama to the Canadian border; five feet blanketed a mountain area of Tennessee.[19] To the east, the center of the storm passed near Philadelphia, where the sun actually came out for a brief period under the storm's fearsome eye. A foot of snow fell on the city, followed by hours of stinging sleet, all of which flash-froze in the early morning hours of Sunday as the back side of the storm dragged frigid air across the East. Philadelphia sidewalks were cemented in impenetrable layers of snow and ice that forced businesses and schools to close for days. A National Weather Service meteorologist described it aptly as an "Arctic landscape."

To the north, the storm ripped Long Island into pieces; 18 homes fell into the water. Later, the island literally was stapled together by the U.S. Army Corps of Engineers. To the northwest, 43 inches crushed Syracuse.

Snow and ice plagued areas far less accustomed to winter assaults. Almost every interstate from Atlanta north was shut down, and every East Coast airport had to cease operations at some point; a record was set for the most weather-related flight cancellations. By NOAA official estimates, it caused almost $10 billion in damages, an all-time high for a winter storm; knocked out power to 10 million people and businesses; and was blamed for more than 250 deaths.

Unquestionably, the toll could have been far worse had the warnings not been communicated so far in advance. It was all but impossible not to know that something huge was coming. The forecasts had underrated the Florida flooding, but for snow and other aspects of the forecast, this was a historic performance for the models and meteorologists, unmistakable evidence of scientific and technological progress.

For the humans and machines to match that performance would be the equivalent of the atmosphere matching the magnitude of the March 1993 superstorm. Nevertheless, they came reasonably close three years later.

The earliest indication of the Blizzard of January 7–8, 1996, bobbed to the surface in computer guidance on the first day of the year. The potential for another life-interrupting snowstorm in the Washington to Boston corridor appeared in National Weather Service discussions on January 3. In this case, however, the computers disagreed even more strongly on the outcome. U.S. models were unimpressed; the European models were. It might have been a case of putting aside national pride: U.S. meteorologists sided with the Europeans. The net result was that the National Weather Service and the private companies sounded the alarms on January 5 and 6. It's a good thing.

What followed was one of the all-time heaviest snowfalls in the East Coast megapolis. Amounts of two to three feet were reported from the Baltimore–Washington area to Boston, and Philadelphia recorded a record 30.7 inches. I had to drive 20 miles to work on the morning of January 7, and about six inches were already on the ground. Never have I encountered such blindingly heavy snows. I was traveling an expressway whose every idiosyncrasy was familiar to me, and a road that could be challenging when it was partly cloudy. Fortunately, I had never witnessed that road so empty. The few of us creeping on the highway used one another's taillights as beacons, like those red-tipped white canes for the blind. I had to make a wild approximation as to the whereabouts of my exit ramp. I was grateful that I guessed correctly, and still more grateful that I did live to tell about it. It was the perfect reportorial prelude to writing about the biggest snowstorm in Philadelphia history. By nightfall, about eight hours after I had parked the car, 19 inches had been reported at the official measuring station at Philadelphia International Airport. As everyone else, I was stranded in the city. My stories appeared in what was then an ultra-limited edition: *Inquirer* online, a new venture. For the first time in its 170-year history, the company was unable to deliver the paper to its half-million subscribers; the governor ordered all roads closed.

As a fitting coda, that night I slept at my sister-in-law's apartment, at her insistence, passing up the company's offer of a hotel room. She lived in Center City about a mile from the paper, and it was still snowing as I trudged through the wind-driven flakes and thigh-high drifts in the utterly snow-barricaded streets. It was almost midnight when I arrived. I had not seen her since the night before, which happened to be her wedding night. Instead of the Caribbean, she and her husband were spending their honeymoon at home—as it turned out on this night, with a brother-in-law. In the morning, it was still snowing. The only place open in town was a ransacked Wawa, where I bought the last two doughnuts to take back to the apartment, one of which appeared to have been half-eaten. It was a rare occasion in which the prestorm supermarket panic was justified. I had witnessed so many of these buying frenzies that I had refused to participate. I did go to the super-market Saturday morning, but all I came back with were six cans of seltzer and a 12-ounce tin of coffee. I argued that we would have plenty to eat at the wedding and should not be tyrannized by a forecast; it wasn't as though the storm was going to keep us from getting around on Sunday. My wife has never forgotten, nor has she let me forget, nor is it clear as to whether I have been forgiven. She has often accused me of underrating a storm's potential because I had a rooting interest in snow. What an outrageous accusation. So what if she was onto something.

The out-of-town wedding guests were similarly stranded, but neither they nor I could blame the forecasters. If any of the guests or any of the 50 million people in the megapolis were unaware of what was coming, they weren't paying attention. This was yet another resounding affirmation of the marvels of emerging computer power and meteorologists' capacity to exploit it intelligently. In its poststorm service assessment, the NOAA, the National Weather Service's parent agency, noted that government and commercial forecasters had done an excellent job of communicating uncertainties, at the same time providing ample advanced warning of a massively disruptive storm.[20]

This was the rare event, almost in a class with March 1993. For the second time in three winters, the meteorological community had executed a forecast tour de force. Had technology and the meteorologists who use it made crucial adjustments to the "new normal," in the popular parlance of climate change, that was yielding a bounty of mega-storms?

If meteorologists were able to score such successes, think how much more accurate forecasts would be with an exponential upgraded in computer capacity. In that context, the "no surprise" comment by John J. Kelly in

January 2000 wasn't all that outrageous. It's just that in retrospect, his timing could not have been much worse, given the stealth storm that blew up a week after his infamous pronouncement. In the 21st century, no storm of that magnitude has blindsided Easterners. In contrast to the Presidents' Day whiff of 1979, the models and their interpreters were well out in front of the Presidents' Day snowstorm of 2003, which set a fresh round of records in the East.

Unfortunately, for the meteorologists, more surprises were coming. In the battle between computers and chaos, the smart money remains on chaos.

"Flizzard" and "Flakegate"

"This will most likely be one of the largest blizzards in the history of New York City." Those were the sobering words of Mayor Bill de Blasio on Sunday, January 25, 2015, coincidentally 15 years to the day from the date of that "no surprise" snowstorm. In the discussion posted at 4:39 p.m. that day, the National Weather Service warned of a "life-threatening historic winter storm expected from late Monday into Tuesday," with blizzard conditions and 20 to 30 inches of snow. "This blizzard has the potential to produce the largest single-storm snowfall in NYC history," the National Weather Service warned.[21]

The mayor took the extraordinary step of clearing the streets of all traffic by 11 p.m. that Sunday. His edict included vehicles making food deliveries, to the chagrin of some of the natives. How did the city expect them to live through this assault by nature with food supply lines cut off? On Monday, at the behest of city officials, businesses closed early, setting off a mass exodus as workers poured out of buildings and crammed into crowded subway cars. Thousands of flights were canceled. New York governor Andrew Cuomo ordered state roads closed. The governors of Pennsylvania and New Jersey declared states of emergency. The megapolis was ready for anything.

Anything but a bust, that is.

By late Monday night, forecasters were shaving their accumulation estimates but by nowhere near enough. Officially, a generic five or so inches were measured in Central Park. It was dubbed the "Flizzard of 2015." On Philadelphia's most popular sports-radio station it was knighted "Flakegate." A winter-storm warning had been issued Sunday with a forecast calling for "14 to 24 inches." The city officially received 1.2. Articulating a question that no doubt has crossed the minds of so many of us, a sports-radio host said of the meteorologists, "How can these people be so bad at what they do?"[22]

What happened? This was a pattern-breaker in that it lacked the prolonged buildup. Unlike in previous cases, the hyperventilation-panic period was compressed. The potential storm didn't make an appearance on the computer-model radar until very late in the game. The accumulation forecasts did not become steroidal until that Sunday, and they were the source of considerable anxiety among the meteorologists. The European Community models were bullish on a monster megapolitan snowfall; the recently upgraded American Global Forecast Model took the low farther out to sea, sparing the cities. On Sunday, the American shorter-range model gave the European some support, and at the end of a 48-hour computer food fight, the American meteorologists cast their votes for the European, as they had in 1996, and which they viewed as superior. One notable holdout was The Weather Channel, which went with more modest amounts, but even those were overdone.

A chastened Gary Szatkowski, the man ultimately responsible for the Philadelphia warnings, took to social media to issue a public apology to emergency managers, airlines, businesses, school superintendents, and anyone else with a Twitter account, presumably even presidential candidate Donald J. Trump.

"You made a lot of tough decisions expecting us to get it right, and we didn't," Szatkowski wrote. "Once again, I'm sorry."

The muffed forecasts set off such a tempest that Louis Uccellini took the unusual step of convening a media teleconference to explain what went awry. I mentioned Szatkowski's apology, and Uccellini emphatically stated he was having none of it. "I understand Gary's frustration," he said, but Uccellini defended the government's performance. He said that his meteorologists had done everything possible within the bounds of the science and that they had not only been right to sound the alarm bells, but also obliged to do so. "Given the uncertainties and the tens of millions of people affected, this was the right forecast to make," he said.[23]

This was yet another instance in which the forecast truly came mighty close to the marks; 24 inches descended upon the eastern end of Long Island. The predicted path of the low was off by 30 to 60 miles. That's barely a pin prick on the planetary surface, but for 30 million or so people in that portion of the megapolis that was targeted in the forecasts, this was the "Flizzard of 2015," "Flakegate," or both.

One person who had no issue with the meteorologists' performance was Philadelphia mayor Michael Nutter.

"I'd rather be wrong and virtually nothing happens," he said, "as opposed to wrong and 14 inches of snow shows up, and we're not ready for it."[24]

Two winters later, Nutter would have the pleasure of a similar experience. That tight margin of error once again was a factor, this time for the 50 million who live and work in the Washington–Boston corridor. In this instance it was a decision on a single word that inflamed the reaction.

Near the end of an uneventful season, on the anniversary of the 1993 superstorm, on March 13, 2017, the National Weather Service in New York hoisted the "B" word, issuing blizzard warnings for just about its entire coverage region, including the most populous areas of North Jersey. The Mount Holly office followed suit with blizzard warnings for suburban areas adjacent to Philadelphia and parts of New Jersey in its forecasting territory. The Philadelphia area had not experienced a pure, verified blizzard since March 13, 1993, although in January 2016, blizzard conditions not far away in south-central Pennsylvania trapped 60 tour buses, 100 passenger cars, and 400 commercial vehicles, forcing the occupants to spend a frigid night on the turnpike. That episode might well have contributed to the invocation of the B word.

The B word certainly attracted public attention. At National Weather Service headquarters, Greg Carbin, head of the forecast operations branch, expressed his reservations about using the word in forecast language. In a conference call with National Weather Service meteorologists in the field, he advised against it. He argued that the storm would be fighting the realities of mid-March. Sea-surface temperatures were warming. They were in the low 40s off Jersey. The strong winds from the east likely would import warmer air that would change the snow crystals to liquid form. This was not March 1993.

"You have to be very careful about pulling out all the stops because it's harder to back off," he told me later in a postmortem interview. "The big challenge is to manage expectations." The warning decisions, however, are in the hands of local offices, and the ones in Mount Holly and Upton, New York, responsible for New York City, went with the bullish predictions.[25]

Carbin was onto something. The storm passed slightly closer to the coast than expected, and for the cities that sabotaged the blizzard potential. In New York, the forecast called for 12 to 24 inches; it got 8. Boston, targeted for 12 to 18, settled for 6.6. Philadelphia got 6, and a bunch of that was sleet. Sleet is liquified snow that refreezes. The snow-melting agent was a stubborn layer of warm air about a half-mile deep that penetrated well inland. Given that it can take crystals 45 minutes to reach the surface, they would have spent several seconds in that warm layer, turning back to ice in the colder air near the surface. That cut down the snow totals. Sleet isn't an efficient accumulator. An inch of precipitation yields a foot of snow or more but only about three inches of sleet. It is dense and presents its own set of hazards, but it doesn't capture public attention like snow.

"How much? How much?" that's all they care about, said Joe Miketta, the acting meteorologist in charge of the office. "Well, not *all* they care about." But it is always *the* question.[26]

Yes, the accumulation forecasts were overdone, Miketta acknowledged, but he pointed out that other aspects of the forecasts, including precipitation amounts, winds, and coastal flooding, were on the money. Ross Dickman, Miketta's counterpart in New York, said the National Weather Service was prudent to hold on to the accumulation forecasts given the track of the storm. "NWS offices decided it was best to remain on the high side of the forecast snowfall in the event the rain/snow line remained just east of the major cities," he said, "which was still a possibility."[27]

Walt Drag, a Mount Holly forecaster who argued for the blizzard warning, did encounter the storm's full blizzard fury. That happened on his way home to northern New Jersey from Mount Holly. He had to abandon his car. Some lesser-populated areas of the state did get their two feet.

New Jersey governor Chris Christie was not impressed. He said the storm was a "big underperformer." And then he offered less than a thank-you note for the government weathermen. "I've had my fill after the last seven and a half years of the National Weather Service, to tell you the truth," he declared.

Szatkowski, at that point retired, fired back. He reminded Christie that the National Weather Service had provided "outstanding" forecasts during Sandy, which made landfall near Atlantic City in October 2012, and generated a catastrophic storm surge along the Jersey and New York coasts. Mount Holly won praise for its PowerPoint-style briefings that cataloged Sandy's potential hazards, a format that became a National Weather Service prototype. He said New Jersey "mishandled" all that information in a "disastrous way." Szatkowski said that Christie's comments after the March 2017 storm were beneath a man who at the time aspired to the presidency. "I'd expect a more responsible perspective from my governor," Szatkowski commented.[28]

Winter storms always have and always will spring surprises. As meteorologists watched the atmosphere make a mockery of the forecast in January 2015, the National Weather Service pleaded for understanding. "Rapidly deepening winter storms are very challenging to predict, specifically their track and how far west the heaviest bands will move," it said in a forecast update. "These bands are nearly impossible to predict until they develop. Our science has come a long way, but there are still many moving parts in the atmosphere, which creates quite the forecast challenge."[29]

What has been different is the welcome scarcity of winter-storm ambushes. An opposite problem has intensified in the third millennium: overwarning. Meteorologists are quick to praise computer models that envision virtual storms before humans possibly could; however, they also are quick to complain about "overmodeling." I attended a prewinter briefing in Mount Holly at which National Weather Service meteorologists boasted that they had access to no fewer than 57 models. They often disagree—both the models and the meteorologists—sometimes dramatically.

"It's called chaos," said Chet Henricksen, Szatkowski's predecessor, now retired. "You look at all these models. . . . It's not like going to the table and betting black or red."[30]

For the foreseeable future, computers and meteorologists will have to co-exist with their limitations.

The public is not fond of uncertainty, but in the end, the only certainty is fallibility. Personally, I am a big fan of fallibility, having much experience with it. I find it reassuring that I have at least one trait to share with the rest of the human race. And I'm not sure I'd want to be part of a "no surprise" atmosphere.

One thing is for certain in the 21st century, if meteorologists believe that a storm is coming—whether they're right or not—the public is going to know it.

Weather has become a national pastime, if not an obsession, and that has a lot do with snow.

CHAPTER SEVEN

Battle of the Titans

The Rise of Commercial Weather

Almost as ubiquitous as the atmosphere itself, weather information in the United States is virtually inescapable. We take this for granted these days, part of the background of our lives, like intrusive music on supermarket and drugstore sound systems. The reality that half the world's population has never experienced snow might be unimaginable to those of who have that privilege on a semiregular or irregular basis.

That there's a 100 percent chance that we'll know if snow is near us (even if it's not) is a relatively modern development. Once upon a time not so long ago, the world of weather forecasting was a dull, if not barren, place, a somber province of staid TV and radio presenters chary with context. The era of saturation weather didn't begin ripening until the late 20th century. That had a lot to do with technology, obviously, and—in my opinion—the evisceration of news-reporting staffs in print, electronic, and online media, the *New York Times* and *Washington Post* being shining exceptions. Weather coverage is an obvious, cheaper, and easier choice than investing time and money in more substantive reporting, and besides it is often visually dramatic. But it also had to do with a handful of visionary figures, people who were passionate about weather and didn't mind if a few dollars fell from the sky in the process.

The late John Coleman, who conceived of the preposterous idea of creating a channel that broadcast nothing but weather 24 hours a day, and Frank Batten, who was crazy enough to buy into it, bear major responsibility for all of this. Another principal figure for me would be one Joel N. Myers. Snow has been a generous contributor to his vast business empire, and fittingly snow had a lot to do with the career path he chose.

A Measure of Injustice

As did several million other people, 13-year-old Joel Myers, a precociously entrepreneurial newspaper delivery boy, surveyed a surreal landscape on November 7, 1953. Myers lived in Northeast Philadelphia, where row houses sprouted like ears of corn in a cornfield. On that day, he lived in an enchanted village. His reaction wasn't necessarily shared by the adult world, but for him it was that dream scenario, reminiscent of what Meta Stern had experienced on the morning of March 13, 1888, in New York. That 1953 freak snowstorm shocked a broad southeast-to-northwest corridor from the Mason–Dixon Line to Lake Ontario. It layered Philadelphia with a heavy, wet snowfall that remains a record for any winter storm that has affected the city before mid-December.

For Myers, the snow was like the ultimate bar mitzvah present. At 13, he was into weather the way the young Mozart was into music. Myers had been keeping a weather diary since he was seven. He was constantly pestering the U.S. Weather Bureau for data. He regularly reported his observations, including snow totals, to the brainy Francis Davis, the weatherman on Channel 6, who was one of the nation's first television meteorologists. Davis would read Myers's reports on the air. Could life get better than that for a child prodigy weather wonk?

On November 7, 1953, Myers went to an intersection near his house, stuck a ruler into the freshly fallen snow, and measured a triumphant 13.0 inches. More than a foot of snow, the biggest snow in his lifetime to date—and winter was still seven weeks away. This was an amazing total for any storm in Philadelphia in that era, let alone one close to Election Day. He was proud of the fact that he had somehow battled nature to load up his bike and deliver the *Philadelphia Evening Bulletin* to the snowed-in subscribers. It was a magical day.

The magic evaporated when the U.S. Weather Bureau reported the official city total from Philadelphia International Airport. It was 8.8 inches.

Myers was enraged. He felt cheated by his government, not unlike the way he would 25 years later. For those who aren't snow geeks, this might be hard to understand: Snow-philes take tremendous pride in their amounts, and even though they take their own diligent measurements, the official totals from the nearest first-order measuring station, one certified and government-controlled for quality and whose totals would be entered into the climate

Cotton Mather, the Puritan firebrand, saw "much rebuke from Heaven" in the colonial snows. *Wikimedia Commons*

Johannes Kepler, author of the witty *Six-Cornered Snowflake*. *Smithsonian Library*

Wilson A. Bentley capturing snowflakes via microphotographs at his Vermont farm. *Frontispiece to his 1931 book* Snow Crystals

Snowflake chart by Wilson Bentley; plate XIX of *Studies among the Snow Crystals. Annual Summary of the Monthly Weather Review, 1902*

Kenneth G. Libbrecht in his snowflake laboratory at the California Institute of Technology. *Anthony R. Wood*

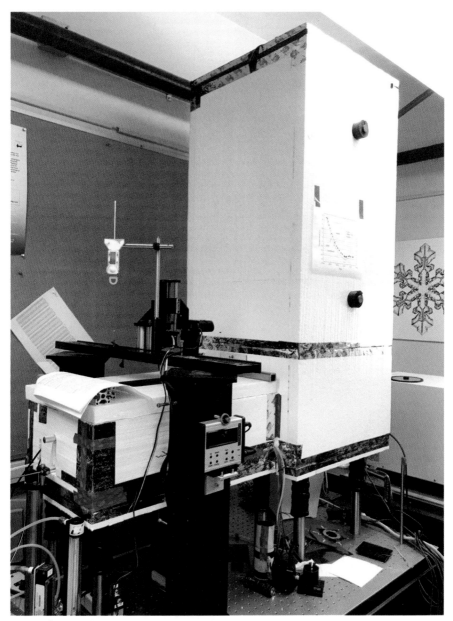

Libbrecht's homemade "convection chamber" for making snowflakes. *Anthony R. Wood*

Snow rollers like those on display in Jackson, New Hampshire, were once used to smooth the snow and spread it evenly over roads. *Anthony R. Wood*

Shoveling out after the Great Blizzard on March 12, 1888. *NOAA National Weather Service Collection*

The Blizzard of 1988 rages in Manhattan. *Library of Congress*

Carting away piles of snow "refuse" after an 1899 storm in New York. *Library of Congress*

Horses and an electrified trolley (right) battle a full-blown storm on January 14, 1910.
Library of Congress

By 1917, cars were trying their luck with the snow, with mixed results. *Library of Congress*

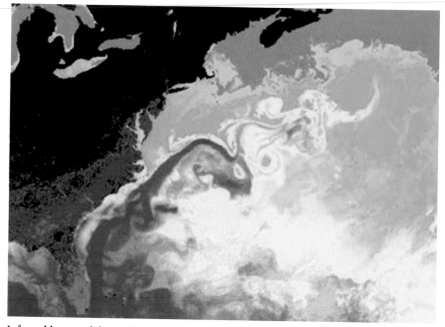

Infrared image of the Gulf Stream current, the fire in the Atlantic. *Wikimedia Commons*

Walt Drag, National Weather Service meteorologist in Mount Holly, New Jersey, work-ing on the forecast on the eve of the March 2017 snowstorm; he argued for hoisting "blizzard" warnings. *Anthony R. Wood*

James E. Church, the Latin professor who figured out how to measure the water in the mountain snowpack. *University of Nevada/U.S. Department of Agriculture*

A mighty white Christmas: Wind-sculpted snow in Glacier Gorge, Rocky Mountain National Park, on December 25, 2009. *John Marino/National Park Service*

Walter Schoenknecht with the general manager of his crowning achievement, the Mount Snow ski resort, in 1968. *Alan Seymour/Mount Snow*

Snow guns going full blast on the bare slopes of Mont-Tremblant in Quebec; even Canadian mountains are sometimes starved for snow. *Anthony R. Wood*

David A. Robinson, luxuriating in a profoundly heavy snowfall in his North Jersey driveway. *Doug Robinson*

A spectacular dendrite snow crystal "manufactured" by Ken Libbrecht. *Ken Libbrecht*

Anthony R. Wood. *Laura Wood*

archive for generations to come, have meanings that transcend logic. Myers demanded justice: What was his government doing to him? Was this some kind of conspiracy? Did the Weather Bureau not have the guts to admit its forecast was so resolutely clueless?

He would get his revenge. That very day, an idea crystallized in his mind that would forever change the business of weather. It was not an original idea. The seed had been planted two years before when his father handed him a newspaper article at the breakfast table. The story was about a woman in Boston who was making money actually *selling* weather forecasts to a local utility company. *Selling* forecasts. To think that someone actually could make money off of something that Joel Myers loved as much as weather. The atmosphere was the province of the government, and the air and everything in it were free. Yet, this woman had found a way to make it rain money.[1]

In the 1950s, young weather geeks like Myers had few options for requiting their passions. The daytime was a weather-information desert. By calling WE 7-1212, one could get a lifeless recording of the Weather Bureau's official forecast from the Bell Telephone Co., purportedly updated hourly. Fellow Philadelphian Elliot Abrams, who one day would be the first of many Penn State students recruited by Myers to work for him, remembered calling that number relentlessly, only to find that the forecasts had not been updated and he couldn't even get a current hourly temperature.[2] How frustrating. As for the radio stations, they simply read what the government gave them, often repeating stale "zone" forecasts. As others, the Philadelphia Weather Bureau office was covering a vast territory. Practically, it would have to update forecasts for multiple counties with radically different topographies. The updates moved via teletype that tapped along at a ponderous 60 words per minute. When the weather updates were needed most, it was all but impossible for them to move at the speed of weather. I had many firsthand encounters with this process when I worked at United Press International (UPI), and one of my duties was to relay those zone forecasts to subscribers.

As for television—and keep in mind that in that precable era, the choices were mightily limited—the weatherman was granted a five-minute segment, including the commercials. The forecasts were delivered sparsely and soberly by mature gentlemen who professed to hate that "white stuff." The Philadelphia market did have one notable exception, "Wally Kinnan the Weatherman," whose name conveniently was pronounced Kin-NAN. Myers would say later that he was an inspiration.

Kinnan was a pipe smoker and affected the avuncular look of his counterparts, but he was quite obviously a renegade. He was a jazz trumpeter, and his segment was introduced by a rather unconventional piano and horn jazz riff, reminiscent of a popular TV detective show of that era, *Peter Gunn*. Kinnan was an anomaly. He was enthusiastic, passionate about the weather, and given to explaining why things were happening and what might be on the horizon. He talked about far-off threats, and in that regard, he was a human precursor of the computer model. He was the first in the Philadelphia market to post an outlandish five-day forecast. Both Myers and Abrams recalled that Kinnan inspired them, adding flesh to the forecasts, a gust of fresh air compared to the Bell Telephone recordings. Sometimes his extended outlooks were even accurate. That's not what mattered; every one talked about the weather but few as interestingly as Wally Kinnan, and certainly no one on television in Philadelphia.

Kinnan and Francis Davis, on-air rivals, cofounded the American Meteorological Society's certification program, which bestowed its seal of approval on those who qualified. It was a badge of honor in the meteorological community. The irony, Davis told me years later, was that Kinnan was not a meteorologist, knowledgeable about the then-esoteric subject of the upper-air patterns that drive weather, which was the source of his five-day outlooks, but not a formally trained meteorologist.[3]

Myers would go on to study meteorology at Pennsylvania State University, eventually earning his doctorate. Penn State would become a national factory for meteorologists, and Myers had a lot to do with that also. He once estimated that he had trained more than 15 percent of the nation's meteorologists. An inordinate number of those were from the Philadelphia region, and that could have been for any number of reasons. Philadelphia holds a prominent place in weather history, and Philadelphians were instrumental in creating the national observation network and a national weather service. My own hypothesis is that the area produced so many meteorologists because the city *isn't* in one of the national hotbeds of extreme weather. For snowstorms, it is in the heart of an atmospheric tease zone. Failure was the norm for snow-lovers, as it was for the region's long-suffering sports fans whose professional baseball team somehow managed to lose a record 10,000 games. As with those infrequent sports titles, frustration made those actual snow hits all the sweeter. Thus, it was only natural to want to know where they came from and what made them happen. For whatever reasons, Philadelphians have generously stocked AccuWeather and other weather outlets throughout the country, Joel Myers being exhibit A.

Myers was a teaching assistant at Penn State in 1962, when a gas company located not far from the central Pennsylvania campus contacted the school looking for someone to provide a winter forecast for energy demands. The head of the meteorology department recommended Myers. *Eureka!*

That story in the Boston paper that his father had shown him when he was 11, the one about the woman making money by selling forecasts to a utility company, had sprung to life. That article was not something Joel Myers would forget.

He volunteered and began forecasting for the company on November 15, 1962. He expanded his venture. He approached ski resorts in the nearby Pocono Mountains and pitched them on targeted forecasts. These wouldn't be the generic government rip-and-read "zone" sleepers: *Cloudy with a chance of snow.* He would tell them what time the snow would fall, how much would accumulate at given elevations, when it would stop, when it would be cold enough to make snow. Here's the deal, he told them: You get 10- and 15-day free trials. If you like them and want more, all you have to do is ask—and pay. He signed up his first ski customer in 1963.

As he pursued his postgraduate degrees, Myers continued to add clients— energy and construction companies, other weather-sensitive businesses. He came up with that name, AccuWeather, by any measure a perfect one. After he acquired his doctorate, rather than pursue an academic or research career, or work for the government, he embraced the entrepreneurial path.

On the Air

In 1971, WARM-AM, a radio station in Wilkes-Barre, close to the Poconos, signed up with AccuWeather, and Myers found himself in the media business.[4] The following year, AccuWeather added the popular all-news radio station in Philadelphia, his hometown, KYW. The other stations were still reading copy from what was now the National Weather Service. KYW had live meteorologists on the air at their beck and call. It wasn't just the forecasts; AccuWeather's specialty was actually explaining the weather, like Wally Kinnan. On snow days, Elliot Abrams's voice on KYW became one of the most recognizable voices in Philadelphia. During his live reports from a windowless radio booth in State College he would talk about what a storm was doing, what it was likely to become, where it was snowing now, and what time it might reach Center City or Northeast Philadelphia or Cherry Hill across the river. He answered questions from the newscaster, ones that listeners might ask; yes, everyone wanted to know "how much?" He would tell them how much "we" could expect. His audience could believe he was in Philadelphia, somewhere down the street.

The company also locked up the Philadelphia TV market in 1972, signing up the most popular news station, WPVI, Channel 6. This wasn't Francis Davis's weather anymore; Davis, a serious scientist, would become the dean of science at Drexel University. The weather segment on the Channel 6 news—now called "Action News"—belonged to a former rock-station radio personality, Jim O'Brien. He was youthful, irreverent, and brash. In terms of meteorology, he was in well over his head, according to longtime Philadelphia TV meteorologist Glenn Schwartz,[5] a Penn State alum who was one of Myers's first employees and who became O'Brien's coach. O'Brien kept the forecasts interesting with a mix of entertainment and science. The science was AccuWeather's; with its expertise, the on-air talent on Channel 6 learned to wax intelligently about the jet stream.

The forecasts were noticeably different from what the other stations were offering, and that was clear to viewers and likely a big reason why the station became dominant in the market. They were different both in form and content. The staid "highs" and "lows" of other TV maps became smiling "good guys" or frowning "bad guys" on Channel 6. But the forecasts were substantive. Rather than a generic call for one to three inches, a snow forecast might call for an inch in Center City and two to three in such elevated areas of the city as Roxborough and Chestnut Hill. Those neighborhoods were 300 and 400 feet higher than downtown. They were away from the "heat island" effect of the buildings and streets of Center City, which absorbed the sun's energy and raised the temperatures. That made all the difference if the readings were near freezing when snow was falling. Plus, temperatures decrease with height, and that would make the elevated areas cooler to begin with. Viewers weren't getting that kind of detail elsewhere, certainly not on the Bell Telephone recordings. Who among the viewers knew this kind of stuff?

AccuWeather was officially building a media empire. It had the number-one radio and TV stations in Myers's hometown. He was just getting started. He replicated the model in cities throughout the country. In each major market he signed exclusive contracts with one radio and one TV client. Among his big scores were WINS in New York and WBZ in Boston, both all-news stations. Broadcasting from a crammed radio booth that looked like a finished basement room turned sideways, Elliot Abrams was on Broadway, he was on Beacon Hill, telling the listeners how much snow "we" could expect. AccuWeather got into the newspaper weather-package business. Its most visible client was USA Today, for which it produced a full-page color map every day with data and narrative forecasts, an unprecedented splash of weather in the staid world of print journalism. Other papers, including my own, liked what they saw and joined the AccuWeather roster. Today it still has 700 newspaper clients.

When an Ally Is the Enemy

Myers was not a beloved figure among rivals, former employees, or government meteorologists. One meteorologist at the National Weather Service headquarters once complained to me that Myers believed the role of government was to help Joel Myers make money.

And it was, and is, true that the profit margins of AccuWeather and other private weather companies have had a powerful ally in the U.S. taxpayer. As the National Weather Service cataloged in a 2017 report on its relations with the private sector, the inventory of largesse includes a

> national set of radars, satellites, and ground stations; the staff who maintain them and provide initial analysis; global agreements through which the NWS aggregates data from around the world; the assimilation of this data into a suite of numerical models; and NWS forecasts that many in the private industry use to varying degree.[6]

The Europeans were known to demand high prices for their weather information. Americans demanded no price. Elliot Abrams, Joel Myers's first employee, has observed that the weather service is a remarkable deal for the taxpayers, giving them fantastic returns on their investments. But ultimately the agency would have what some in the private sector viewed as a deleterious effect, and relations between AccuWeather and the National Weather Service often took stormy turns.

By 1978, AccuWeather ruled in Philadelphia, but it had a major slipup in January in Myers's hometown. Its forecast for January 19, read and repeated all night on KYW, called for one to three inches of snow changing to rain, as had been the pattern established by previous storms that month. KYW repeated the one- to three-inch forecast several times each hour. It was similar to the National Weather Service forecast. I was working the late shift at UPI that night and transmitting that forecast to local radio stations. Like the National Weather Service and evidently the meteorologists in State College, I was unaware of what was going on outside. When I left at midnight, I had to walk home in knee-high drifts. Philadelphia got creamed. More than 13 inches were measured officially at the airport in what had turned out to be one of the worst storms on record for East Coast cities. It was an ugly night for drivers and forecasters.

AccuWeather and the government recovered nicely two weeks later, well out in front of another nor'easter. AccuWeather boldly predicted that it had a chance to smash the all-time record in Philadelphia, which at the time was 21 inches. The National Weather Service forecast was on board with a storm forecast, but its accumulation estimate was far more conservative, on the order of six inches "or more." When the snow stopped on February 7, it had set a record in Boston—30-plus inches—but not in Philly. For the UPI story, I spoke with Myers and informed him that officially the airport had measured 14.1 inches. He expressed outrage. In November 1953, the government had cheated him by more than four inches. Now this. I asked him if he thought perhaps the government was undercounting the snow because it had gone so light on the forecast. Myers didn't dispute that conclusion.[7]

A year later, when the surprise Presidents' Day Storm of 1979 imposed a government shutdown on the nation's capital with two feet of snow, Myers ripped the National Weather Service for blowing the forecast. He said the agency was relying too heavily on computer modeling, which whiffed on the fact that the storm was bombing out off the coast, instead of paying attention to how the actual weather was unfolding. Myers said that his forecasters had done just that and that its accumulation estimates were closer to reality than the government's. Chet Henricksen, who was the government's lead fore-caster at the height of the storm, argued that in reality, the opposite was true.[8]

I saw Joel Myers in person for the first time at a conference in Washington, DC, in 1983. About a dozen media representatives were invited to a discussion of "What We Know, and Don't Know about Weather." He was among the experts—including several high-powered weather service officials like Neil Frank, then the head of the National Hurricane Center; Fred Otsby, who ran the tornado-forecast center; and Don Gilman, in charge of the seasonal outlooks—invited to address the assembled media. All the speakers offered what to me were fascinating insights about the state of the science. Myers's talk was essentially a business pitch. He passed around samples of his company's newspaper packages, perhaps unaware that a majority of the people in the room did not work for newspapers and that those of us who did were in no position to make purchasing decisions for their companies.

In the spirit of the conference topic, when he was finished I asked him what AccuWeather knew that the National Weather Service didn't. He allowed that it was a fair question but then pivoted from gentle salesman to dissatisfied customer. His complaint was with one of the products of the government's largesse: He said the National Weather Service's most recent version of a model that meteorologists were relying upon for predicting storm development was making forecasts worse, perhaps with the model's perfor-mance in February 1979 in mind. A National Weather Service meteorologist in the room was nonplussed, obviously not anticipating such an aggressive comment from a fellow presenter and from someone whose business was reli-ant on his agency's services. The discussion did not degenerate, but Myers's comment had a clear edginess, and it was a glimpse into a relationship with the National Weather Service that would become frostier throughout the years—and worsen with the development of the internet.

One of the ironies for the commercial companies has been that the tax-payers' generosity also has been a bane of their profitability, in their views. That is a subject that Myers has taken to Washington. In 1998, he publicly asked Congress to order the agency to stay out of the routine forecast business. No, that did not happen.

But even Myers's rivals and detractors in the public and private sectors acknowledge that he is a giant in the industry. One of them described him as the Bill Gates of weather. From its giveaway mom-and-pop origins, AccuWeather has become an international behemoth, a dominant force on the web and a go-to place during snow events. It operates out of a colossal, modern headquarters on the edge of State College. Myers has been a business owner and chief operating officer, and he has also been involved in day-to-day forecasts, done broadcast feeds, and served as an in-house scientist. He is author of what one of his well-known hires, Joe Bastardi, has described as the "rubber band" theory, which holds that when the atmosphere locks into an extreme state it will snap back and change abruptly rather than gradually. That's why it often happens that spells of frequent snow and cold are routed by periods of unseasonable warmth.

If weather had a Hall of Fame, Myers would be a first-ballot selection. Could he have become an even larger figure and AccuWeather even more dominant?

TV or Not TV?

Although to this day he says that he has no regrets, Myers would watch another company exploit the rapidly growing universe of cable television and become a titanic rival.

One of the media figures attending that Washington conference in 1983 was a gentleman named John Coleman, a TV legend who was the first meteorologist on *Good Morning America*. In his candid business memoir, cable entrepreneur Frank Batten recalled that in the early 1980s, he was on the hunt for the next hot concept in programming.

How about an all-news channel? No, that was out of the question because a sailing friend named Ted Turner had beat him to it with CNN. Then a preposterous possibility presented itself. A venture capitalist approached Batten's chief financial officer and said his poker partner had a business plan, a concept that was certifiably outside the box, well before that phrase became

popular. The poker partner was John Coleman.[9] His idea was to start a channel that would do nothing but broadcast weather 24 hours a day. Coleman,
based in Chicago, although viewers assumed he was in New York on the
set with host David Hartman, would work all night on his *Good Morning
America* stints with meteorologist Joe D'Aleo. After work they would eat
breakfast and talk about Coleman's dream. Coleman saw the nation as
weather-deprived, merely teased by what was available on television. Whenever a promising backer surfaced, Coleman would think nothing of hopping
a plane for a meeting. He found a receptive audience in Bill Diederich,
Batten's CFO, and, eventually, the boss himself. Batten, a Harvard MBA,
was the successful publisher of the *Norfolk Virginian-Pilot*, which, in 1960,
won a Pulitzer Prize for its bold editorials on school segregation. The paper
was owned by the Batten family's Landmark Communications, Inc., which
entered the fledgling cable business in the 1960s.

AccuWeather was started by an entrepreneur who was a trained atmospheric scientist. Batten was no scientist, but he most certainly was an
entrepreneur who professed to be "fascinated" by weather. He was a sailor,
for one thing. And at age six, he had a dramatic and memorable weather
experience. He was staying at his uncle's cottage at Virginia Beach when it
was ravaged by the devastating "Potomac–Chesapeake Hurricane" in August
1933. He was drawn to Coleman's idea. After intensive negotiations, analyses, and background checks on Coleman, Landmark bought into Coleman's
concept.[10]

Was Batten's decision hailed as brilliant? Is it sunny on the dark side of
the moon? He said he shopped the idea to a lot of venture capitalists, who
expressed no interest. Batten said he had "several" reasons to believe it would
work, including the testimony of his marketing expert, who was in charge of
the TeleCable subsidiary. TeleCable was offering channels with nothing but
basic instrument dials and weather-forecast texts. But Batten recalled him
saying, "Whenever the power went out we got more complaints from people
about the loss of that primitive weather channel than about the programming. So that gave us a lot of confidence that this was something in high
demand."[11]

Still, 24 hours of *weather*? This was the equivalent of a 24-hour test pattern. True, when WINS in New York and KYW in Philadelphia went with
the 24-hour news format, those had seemed like forays into lunacy, lunacy
rooted in monotony. Yet, they became dominant in their markets. In the
court of common sense, however, The Weather Channel looked like a bridge
too far.

Batten recalled convening a news conference at the Park Lane Hotel in Manhattan on July 30, 1981, a time when the media world was abuzz with ambitious proposals to exploit the cable market. About a quarter of U.S. households had cable at the time. The room was crowded with reporters from the major media outlets and trade journals. If nothing else, Batten said, they did get a surprise. "I am here today to announce an exciting new concept in cable television," Batten told them. His company would be launching a radical new venture and be on the air the next year. "It will be all weather, 24 hours a day. We'll call it . . . The Weather Channel."

The reaction: "First silence, then a collective groan."[12]

Batten plowed ahead anyway, and The Weather Channel (TWC) went on the air on May 2, 1982, with Coleman in charge of operations. It was not an instant hit. To complicate matters, Batten and Coleman had an unsightly and public falling out that ended up in court. Nevertheless, in his corporate memoir Batten acknowledged, "The Weather Channel was John Coleman's idea and his dream. His energy, drive, and determination to get it started were remarkable." Coleman was a visionary, but Batten described Coleman as temperamental and a poor manager who lacked business acumen. Under Coleman's leadership, TWC was losing $1 million a month—the equivalent of $2.7 million in 2019 dollars. "We were literally hemorrhaging red ink," Batten lamented.[13] TWC was collecting no fees from cable systems, and wasn't selling enough advertising to cover the losses. Coleman filed a lawsuit that drew public attention to TWC's struggles. Batten decided that he would let the channel go dark. But cable operators implored Batten to keep it alive.

Why? It wasn't as though TWC was as popular as, say, *Monday Night Football*. "A number of other cable programming services were dying," Batten said. "CBS had a cultural channel, and that had closed. ABC had a news channel, and that was about to close, and the cable operator didn't want to lose any more programming." Batten said TWC would stay on the air only if cable operators were willing to pay fees. He gave them a three-month deadline to decide. He let them see for themselves TWC's financial troubles. "We opened our books to the cable operators, and they realized from that we really were on our last legs."

The appeal worked. "Within three months about 80 percent of the cable operators had agreed to pay us fees," Batten said. TWC survived, to say the very least.[14]

With a huge assistance from taxpayer-captured data—the fuel driving private-sector weather—it generated monster profit margins, in its heyday up

50 percent. In its early years, the on-air personalities were foregrounds to the weather. Thus, TWC wasn't paying network anchor salaries. The channel didn't even provide local forecasts: Those came from the National Weather Service, the agency's printouts posted on the screen as a bridge to the commercial breaks. If AccuWeather's franchise was micro-weather, TWC's was macro-weather, a TV version of *USA Today*. Whatever it was, it worked.

Myers told me a few years ago that he was at peace with his decision not to follow Batten's lead and jump into the cable market.[15] In the early 1980s, he said, he wasn't prepared to make the capital investment. Besides, he had done wonderfully for himself. By his reckoning, among the hundreds of commercial forecast services, he had one of the most popular weather companies in the world. AccuWeather claims 1.5 billion users worldwide (up from 600 million in 2012). Since AccuWeather is chary with its business secrets, that would be impossible to verify, although its reach is indisputable. A similar calculation for The Weather Company, the parent of The Weather Channel, weather.com, and other entities, which it has swallowed whole—all later acquired by IBM, which has become a weather-business behemoth—would be even more complicated. In a 2018 survey, accuweather.com was far and away the number-one weather website, with about half of the visitors from outside the United States. That was more than twice as many as weather.com and its affiliated sites.[16]

But in terms of raw dollars, the crown belonged to The Weather Company, which generated about twice as much revenue as AccuWeather, according to a 2017 analysis by the National Weather Service. It estimated that all private services harvested about $7 billion worth of business annually. AccuWeather and The Weather Company were the giants, accounting for 30 to 50 percent of that total.[17]

$now

Taxpayer generosity certainly helped make all that possible. But those titans also owe an immeasurable debt to snow.

For luring a captive audience, "snow" is the magic word. If snow stops a region cold, all the better for the commercial companies who are ideally situated to prosper from the traffic from the shut-ins. Snow stokes the search engines, and from the beginning for AccuWeather and TWC, it stoked the cash engines. At the AccuWeather micro-level, when it was snowing in Philly, KYW and WPVI owned the franchise; KYW and WPVI had AccuWeather. On the macro side, For The Weather Channel, the beauty of winter was that it was almost always snowing somewhere.

AccuWeather was instrumental in redefining the business of weather; TWC has been instrumental in redefining the public's perception of weather. A blizzard in New York's Snow Belt could become everyone's blizzard. When TWC sent a reporter to stand in the wind-frenzied falling snow, a reasonable person might ask, "Why is that poor person out there freezing?" The better question might well be, "Why are people watching this poor person who is standing out there freezing?" This is the lazy man's way of looking out the nation's window at the winter fury without the skidding tires, scrapers, and shovels. If it was snowing in Traverse City, Michigan, the nation could share it.

When Achilles Met Andrew

Like winter storms, hurricanes are superstars of the atmosphere among Americans, fantastic audience-drivers for the weather companies. They are photogenic, their fearful symmetry spinning menacingly across the Atlantic in slow-motion satellite imagery. They are ponderous. They can take several days to make the trip from the Cape Verde islands off of Western Africa to the populous U.S. East Coast. That is more than ample opportunity to build the drama. A major hurricane becomes the long national horror movie as the monster storm approaches the built-out coastal states. And ratcheting up the suspense is the fact that forecasters, not even the brightest minds in tropical-storm research, can never be absolutely certain where the monster is going to land until it actually does. Small differences in path can make gigantic differences in impact. Hurricane Andrew, for example, as horrifying and damaging as it was, would have been far more destructive if, after its 3,000-mile trip across the Atlantic, it had made landfall 40 miles farther north, in downtown Miami, rather than Homestead.

Yet, as widespread as their impacts can be, in geographic coverage most hurricanes are comparatively small. In a larger storm, hurricane-force winds might extend to a 150-mile radius from the center. Winter storms, by contrast, can be triple the size of hurricanes in areal coverage. While the maximum impact zones of hurricanes are tightly focused, a winter storm can have maximum impacts on multiple states. That March 1993 nor'easter affected a third of the nation's population.

But for drawing viewers, listeners, and clickers, whatever they lacked in geographic coverage, hurricanes historically held a distinct marketing advantage over winter storms. They had *names*—for practical reasons. As has occurred in any number of seasons, two or three tropical storms have swirled in the Atlantic Basin at the same time, and identifying them by names helps keep them ordered in the public mind.

Beyond any practical considerations for affixing names, hurricanes are ripe for personification.

Like humans, hurricanes have "eyes." Like humans, they often engage in erratic behaviors. Like human invading forces, they make "landfalls."

In 2012, TWC decided to extend the limits of personification while attempting to level the marketing playing field between those two superstars of the atmosphere. It started naming winter storms. And that act set off a storm that deserved a name in and of itself.

TWC held that it was motivated by concerns for public safety. Naming winter storms would make everything so much simpler for emergency management types and everyone else. Winter-storm names were perfect in the 21st century, natural Twitter hashtags. "Coordination and information-sharing should improve between government organizations, as well as the media, leading to less ambiguity and confusion when assessing big storms that affect multiple states," it said in announcing its initiative.[18]

Joel Myers and AccuWeather took immediate exception. This was war, a brazen attempt by a competitor to hijack a coveted cash cow—the snow. The company fired off a news release. "A unilateral decision by The Weather Channel to name winter storms will create confusion, rather than delivering critical and important safety and planning information to the public," it said. It quoted Myers as saying, "In unilaterally deciding to name winter storms, The Weather Channel has confused media spin with science and public safety, and is doing a disservice to the field of meteorology and public service."

"Myers is probably yelling at people for doing it first," groused one Philadelphia meteorologist who was an AccuWeather alum.[19]

Not so, Myers insisted.

"We have explored this issue for 20 years and have found that this is not good science and, importantly, will actually mislead the public. Winter storms are very different from hurricanes."

That point is inarguably true, and not just in terms of size. Winter storms are not singular events whose progress can be monitored for days. An East Coast storm, for example, often starts with a disturbance in the Pacific that becomes a weak Alberta Clipper moving across the United States from southwestern Canada. It might get shredded by the Appalachians, only to mutate into a major nor'easter when it encounters the Gulf Stream–influenced Atlantic Ocean. AccuWeather argued, correctly, that some of the most severe winter-storm impacts are highly localized. What about those lake-effect blizzards? Should they be named? And what of those renegade ice storms that put towns and cities on skates?

In short, AccuWeather argued, TWC was compromising public safety for the sake of profit.[20]

Then there is the naming process itself. The naming of hurricanes is based on strict criteria. A storm must be tropical in character and have maximum peaked sustained winds of at least 74 miles per hour, verified by aircraft observation and subject to rigid quality controls. TWC's names are based on the forecast, rather than based on observation. TWC holds that nonetheless, this is not a frivolous concept. Winter-storm specialist Tom Niziol, a highly regarded meteorologist, explained that the naming committee looks for quite specific criteria: a storm that would affect at least 2 million people—about the combined population of Manhattan and Staten Island—and/or cover an area of 240,000 square miles, an area a tad smaller than Texas. In addition, the committee has an elastic clause, Niziol wrote.

> The storm-naming committee still reserves the right to override the quantitative decision in certain circumstances. Some of the factors that may influence decisions to override the naming rules include the degree of historical significance of the event (e.g., accumulating snow in South Florida, a summer season snowstorm, etc.).[21]

The concept has support in the United Kingdom, as Mary M. Glackin, The Weather Company's senior vice president of "public–private partnerships," pointed out. The U.K. Met Office names its cool-season storms. While true, it is quite different from the TWC model, and in more than names only. TWC's 2018 list began with Avery, Bruce, and Carter; the U.K. Met Office's first three were Ali, Bronagh, and Calium. TWC's names apply only to storms with frozen precipitation, but the winter-storm climate of the British Isles is very different from that of the United States, and the U.K. Met names can apply to rainstorms.

The Met Office says the naming has been successful. It says that 80 percent of people surveyed found the system "useful for making them aware that the storm may have greater impacts than normal, with 63 percent agreeing that storm-naming was useful to let them know to take action and prepare for the impacts."

So why, Glackin asked, can't we all get together, sing "Kumbaya," and play nice. "What's important is to lead as a community in this social media era," she wrote. "During major snow events, the reach on Twitter has been over a billion. What would our reach be with all of us working together? . . . Are we ready to reengage on this topic as a community?"[22]

The odds of getting Joel Myers to "reengage" are right down there with a blizzard hitting Miami.

The biggest difference between the U.K. Met naming system and TWC's is that the former is a public, government initiative; the latter is an initiative by a profit-making enterprise. It is unlikely that AccuWeather or any of the surviving competitors in the private sector would participate in anything under the TWC umbrella. The National Weather Service officially is keeping out of it.

Glackin argued that the value of storm-naming transcends competition. "What's in a name?" she asks. "Well in this case, the name is the headline to attract attention to the threat. . . . In this information-saturated world, this headline/hashtag is key."[23]

The concept has had some support in the meteorological community beyond TWC. Jason Samenow, the meteorologist who is the chief of the *Washington Post*'s Capital Weather Gang, said it could have value. Glenn Schwartz, the dean of Philadelphia meteorologists, likes it, and for a period his station experimented with naming storms. As Samenow pointed out, however, the source in this case is problematical.

"I would have preferred the National Weather Service take this on rather than TWC. An NWS storm-naming initiative would have more credibility since its mission is to protect life and property rather than to make money and generate publicity," he wrote.[24]

As to whether the naming has had an impact, that might well be impossible to quantify, that U.K. Met study notwithstanding. In the only disinterested U.S. study that I'm aware of, a group of researchers at the University of Connecticut found no evidence that it made any difference in terms of the public's response to a given threat. It involved 400 subjects and included that paragon of 21st-century truth, Twitter. The researchers presented the subjects, broken into three groups, with three mock Tweets reading, "Up to 1 FOOT of #SNOW for parts of New England."

In group one, the winter-storm name "Bill," typical for a tropical storm, was included in the Tweet; for the second, it was a more Weather Channel–like "Zelus," and for the third, the storm was nameless.

In the study, published in the journal *Weather, Climate, and Society*, whether it was "Bill," "Zelus," or no name made negligible differences in terms of the participants' perceptions of the storm's potential severity.[25]

Regardless of the merits of the case for naming storms, unless IBM somehow swallows AccuWeather and all its remaining competitors and manages to annex the National Weather Service, it will be naming storms on its own. Snow is simply too valuable a forecast commodity. For example, in February 2010, a profoundly snowy one from Washington to New England, Accu-Weather reported more than 400 million page-views in just that one month, beating its former monthly record by 60 million. On one day, February 10, when the second major snowstorm in five days shut down almost all traffic in the East except for web traffic, accuweather.com logged 21 million visitors.[26]

These days, with weather information available on so many varied platforms and social media, with so much of it cycled and recycled by secondary sources, it would be impossible to tease out just how much people rely on the government versus the private sector for what it wants or chooses to know about the weather.

One thing is known: The private sector relies heavily on the government. When she was head of the National Oceanic and Atmospheric Administration (NOAA), Jane Lubchenco quoted a member of Congress as saying, "I don't need NOAA's weather satellites; I have The Weather Channel." I guess you can fool at least one congressperson some of the time. Without the taxpayer-funded NASA, NOAA, and the National Weather Service, there would be no TWC, no AccuWeather. The government has been private companies' biggest benefactor, and unresentfully so.

For its part, the government and the taxpayers get the world's most comprehensive weather-information distribution system. The United States is saturated with weather.

For their parts, the commercial enterprises are free to accept what their government offers, add value, and then market their refined products to their customers. They can customize snow forecasts for highway departments or utilities, giving them specific starting times in mile-marker areas and hour-by-hour forecasts, and be on call for consultation during storm events.

The commercial services defer to the National Weather Service on official severe-weather advisories. TWC and AccuWeather dutifully post government-issued winter-storm warnings, even when the companies' forecast wording clearly suggests that they don't believe the snow will reach the warning criteria. The National Weather Service is the point of contact for county and state emergency managers before and during winter storms.

Joel Myers and his brother Barry, CEO of AccuWeather, would not mind at all if all the National Weather Service did during the winter was issue winter-storm warnings. That sentiment blew up into a public spat in 2005, and resurfaced with a vengeance after President Donald Trump was inaugurated and the government shut down in the winter of 2019.

At issue is the fundamental role of the National Weather Service. AccuWeather has been at the forefront in complaining that the government is way too generous with its data, forecasts, and products. In effect, Myers has argued, it is giving away the store.

These days, National Weather Service sites throughout the country do offer a treasury of information that would have been unimaginable to the young Joel Myers and Elliot Abrams. This isn't the Bell Telephone Co. recording anymore.

For a snow threat, the National Weather Service posts quite specific accumulation maps. And not just one for overall totals. It also posts them for potential low-end and high-end amounts. It offers probability tables town by town. For Philadelphia, the column for four inches might show a probability of 30 percent, and for six inches, 20 percent; for Wilmington, 15 miles away, the figures might be 20 percent and 10 percent.

The areal forecast maps that pop up on their sites have a "point-and-click" function. Put the cursor near where you live or work, or where you want to be, and get a forecast. Or if you care, plug in your town in the search function. Want the snow probabilities hour-by-hour for a given location? Go to the "tabular" output and you'll get the hourly readout for temperature, dewpoint, humidity, winds, windchills, precipitation chances, and about everything else outside of your odds for winning the lottery. If all you care about is snow, you can filter out the other options.

For the public, it's quite the buffet of freebies; for Myers, it is unfair competition that is hurting his business and the entire private weather sector. He has done more than complain.

A Capitol Idea

The bill was called the "National Weather Service Duties and Responsibilities Act of 2005." As the title suggested, its aim was to "clarify the duties and responsibilities of the National Oceanic and Atmospheric Administration and the National Weather Service." Essentially, the bill called for those duties and responsibilities to be as minimal as possible. It was introduced by Senator Rick Santorum, a Pennsylvania Republican, and it could not have been more favorable to the private sector if Joel Myers, himself, had crafted the language. Some are certain that in fact he had, or at least that he had a large hand in it. Judging from the specificity of the bill, not to mention the embedded jargon, it is certain that Santorum didn't.

Myers and Santorum had not only a Pennsylvania bond, but also that Happy Valley "We are Penn State!" connection. State College is truly a college town, insular and lovely in the geographic center of Pennsylvania; AccuWeather has become a Penn State annex. Joe Bastardi, who was an AccuWeather web legend and hero among his "snow geese" for his daily columns that frequently mentioned snow threats that no one else was seeing, declared a number of times how proud he was that late Penn State football coach Joe Paterno would greet him and his son by name when they saw him around town. Joel Myers, then a member of the Penn State Board of Trustees, was among those who publicly expressed outrage at the horrific sex scandal involving former football coach Jerry Sandusky, who went to jail for sexually assaulting young boys in the storied football locker-room shower. Joel Myers has contributed generously to his alma mater, and the elaborate university weather center bears his name. The Myers brothers also contributed generously to the campaigns of Penn State alum Rick Santorum.

Santorum's bill would have relegated the role of the National Weather Service to issuing those aforementioned warnings, while continuing to provide all that observational data and other products essential to forecasting to the private companies. *Bloomberg Businessweek* has quoted Santorum as saying that one of the Myers brothers came to his office to complain about the National Weather Service in 2005, before he introduced the bill. The role the brothers had in the bill's wording was unclear, but they would not have complained about it.[27]

In the section entitled "Competition with Private Sector," the National Weather Service would have been enjoined from "providing a product or service that is or could be provided by the private sector" unless the secretary of commerce "determines that the private sector is unwilling or unable to provide such product or service." So much for those point-and-click forecasts and snow-probability tables. During a snow threat, TWC, AccuWeather, and other private companies stood to reap even larger audiences.

The bill set off a tempest in the meteorological community.

Cliff Mass, a prominent atmospheric scientist and climate researcher at the University of Washington, was among those who felt compelled to come to the National Weather Service's defense and speak out against the bill.

"It would very seriously limit what the Weather Service does," he said.[28] He likened it to barring the U.S. Postal Service from delivering the mail; think what a boon that would be to Federal Express and UPS. The bill generated such opposition that it didn't draw a single cosponsor.

It died, unlike the tensions between the private sector and the government, which flared to a new level of intensity after the presidential election of 2016.

When President Trump went looking for a new head for NOAA, the National Weather Service's uber-parent, he found his man in State College. Ten months after taking office, Trump nominated Barry Lee Myers. One consequence of the nomination might have been a record outbreak of "storm" references in media headlines. The National Weather Service

Employees Organization, which represents the agency's meteorologists, was apoplectic. Union officials complained that Myers was an attorney, not a scientist, that he knew nothing about the oceans, or satellites, or atmospheric research, things that constituted the bulk of NOAA's budget. But their overriding concern was that they foresaw the end of free weather, a not-necessarily bloodless coup aimed at transforming a taxpayer-funded "service" into a business, one so deeply subsidized by consumers that in effect, they would be forced to pay twice for their weather.

"Myers's nomination would present a host of conflicts of interests," the union argued. "As NOAA administrator, he would be in a position to fundamentally alter the nature of weather services that NOAA provides the nation. If approved, he will be able to order the National Weather Service to do precisely what his company was unable to accomplish through legislation."[29]

The record-long government shutdown in the winter of 2018–2019 was a traumatic period for government meteorologists, who were forced to work but didn't receive paychecks. Like the people they served, they had families, mortgages, and other expenses that come with existence. The partial shutdown began right before Christmas, and the impasse dragged on through January as President Trump demanded that Congress appropriate money for his proposed wall along the Mexican border, and the Democratic majority said hell no.

For the National Weather Service union, however, the shutdown did have one positive consequence: It provided more ammunition to aim at the Barry Myers nomination.

The headline on the AccuWeather news release on January 17 read, "Can you trust weather forecasts during the government shutdown?" It cited weather havoc on the California coast and a major-snowstorm threat in the Northeast. It quoted media reports as saying, "The limitations on the National Weather Service will make forecasts 'worse,' and that presents a 'national security risk.'"[30]

But not to worry, it said, "Not when there are alternative weather forecasting sources."

The outrage from the weather community came rapidly and furiously. "Talk about kicking someone when they're down," said Jim Eberwine, a veteran National Weather Service and AccuWeather alum.[31]

AccuWeather ended up issuing a public apology for the release, but the National Weather Service union had its opening. It argued that the insidious implications of the release were further evidence of AccuWeather's grand scheme to annex the government, that the nomination of Barry Myers would do nothing but further those ambitions. Barry Myers agreed to sever his connections with the company. The union's response in essence was, "Can you stop being Joel's brother?"

During the shutdown I frequently consulted National Weather Service sites and interviewed meteorologists, and I would say from my perspective that the shutdown had absolutely no effect on the quality of forecasting or output of products. I was in charge of the team covering the local impacts of the shutdown, and we found no grist from the performance of the National Weather Service.

The irony is that the fact that some government employees continued to work without pay prolonged the shutdown. Dan Sobien, head of the National Weather Service union, captured the paradox: "If the government ever really shut down, we would never have another shutdown."[32] In that notorious release, Myers is quoted as saying, "AccuWeather brings more weather data and weather models into our facility than any other place on the planet." But, yes, he said, that did include U.S. models and government-captured data.

The confirmation skirmishes raged all the way into the fall of 2019. The union and other critics ultimately got their way. In November, Myers withdrew. Myers said that his nomination was doomed by his loyalty to President Trump and that his family had been "mercilessly attacked" by "false" news stories. But he insisted that none of that dissuaded him from pursuing the appointment: He withdrew for health reasons, having recently undergone cancer surgery. The Senate never got around to voting on his nomination.[33]

The Enemy Within

Ironically, the private companies have been struggling with a competition far more complicated and challenging than any competition with the government or their commercial rivals. In the last 25 years, they have been competing with themselves. While AccuWeather has complained about the government candy store, it also has been offering information free of charge to anyone willing to endure aggressive popup ads for various things you probably didn't know you wanted. With the surge of the internet's popularity in the 1990s, the commercial outfits were presented with new opportunities, and new dilemmas, as were other businesses, including my own.

Newspapers were unprepared for the communications revolution, and by the time they decided how to reinvent their strategies, their businesses were permanently damaged. My paper and others decided in the 1990s that they had no choice: Content for which print costumers would be paying would be offered free online. Not only were people finding it more convenient to go online to shop for what they wanted to know, but also they could get some of the more deeply reported and specialized stories from their own newspapers for nothing. The predictable result was an entitlement mentality, as in why should anyone pay for it? Yet, newspaper companies felt they had no alternative. If they didn't have an online presence, they might evaporate from the public consciousness altogether. Our company created philly.com, separate from the newspapers the *Philadelphia Inquirer* and *Philadelphia Daily News*, but its content came primarily from the reporters and editors of the *Inquirer* and *Daily News*. Plus, we were giving our online readers the advantage of not having to wait a day for the stories. Reporters and editors put up a wall of resistance: We were posting stories before they were completed and tipping off our competitors.

In the old days, when it snowed, we spent the day reporting and gathering evidence of trauma and anecdotes. It was usually my job to make the "sausage," as I called it; that was the overall roundup story incorporating all the various feeds from reporters. One of my favorite *New Yorker* cartoons depicted a sausage at a job interview. The sausage says, "My background, you don't want to know." Typically, I would file the story about 7 p.m., and we would update as warranted for the later print editions.

On the morning of January 8, 1996, in addition to the weather making history, the *Inquirer* made the kind of history it didn't want. For the first time in its storied existence dating to the 1840s, it was unable to deliver the papers. The governor had ordered all roads closed because of the record blizzard, and that meant that our trucks couldn't leave their bays. I learned days later from a friend that her sister had read my stories that day: They had appeared on philly.com. If I ever knew we had such a thing, I had forgotten.

That, of course, has changed. We are now a "digital first" company, and nothing drives readership like the word "snow." It doesn't have to be actual snow, although that helps.

On snow days, our online traffic zooms. For that matter, it can zoom even when it isn't snowing. A forecast is enough. I remember our site getting brisk traffic on a story that it might snow in the Pocono Mountains, a ski area in upstate Pennsylvania—sometime next week. I am proud to say I did not write it. The word "snow" in a headline is a pure click magnet. An item about a virtual storm that existed only as a gleam in a computer model's eye could draw monster traffic. I avoided those also.

When snow actually is falling, it is all hands on deck all the time. The story becomes a work in progress. We no longer have the luxury of waiting until day's end to make the sausage. Readers now get to watch the sausage-making progress, with a running narrative updated frequently with such "need to know" tidbits as school closings, business cancellations, and flight delays. Yes, everyone might be home watching television, but evidently that appetite to know about the snow is insatiable. They also are on their laptop and desktop keyboards. On snow days they can't get enough of us, so we were accommodating our visitors with all the free information that money need not buy.

At one point, we did attempt to set up a hard paywall for the newspaper content. It was ill-fated, however, because at the same time we were offering almost all those articles gratis on philly.com. Imagine walking into a restaurant and being told that you could eat for free downstairs but have to pay upstairs. Where do you think you would eat? In the end, we returned to the "free concert" model and, like other media organizations, evolved to essentially a free-trial system. If you get to a certain number of articles the subscription wall goes up, but the industry continues to fight that entitlement mentality. Some folks simply don't want to pay for our hard work; it's nothing personal.

Weather companies experienced something similar. It would be borderline inconceivable to anyone who hasn't spent at least a few decades on the planet to accept that when TWC first went on the air in 1982, no weather companies had websites, nor did the National Weather Service. Like the newspaper business, once the internet revolution started, the private weather companies had no choice but to offer their forecasts for free to online consumers. They had no choice but to compete with themselves. This created complications for two of the titans of the commercial weather market.

TWC was on the air more than a dozen years before its internet affiliate went live in 1995. And it wasn't as though the company knew what it was getting into. Just a few years before it had acquired the domain name weather.com, not having a clue about what it might do with it. "When we launched weather.com, the internet was still a clunky, unformed, poorly understood medium," Batten said.[34] His son, head of new ventures for Landmark, had warned that the internet was a "disruptive" technology that would change the way the company did its business, with a particular impact on TWC. Would the online enterprise draw eyes from TV and compete with it for advertising dollars? Batten said that in the company's internal discussions about the future of weather.com, the "key players" agreed that "we wouldn't be driven by concerns that it might compete with The Weather Channel, or cannibalize the audience." In-house tensions were unavoidable, however. TV staff complained that upstart weather.com was getting favored treatment in hiring and balked at promoting weather.com on the air. Ultimately,

Batten decided to make weather.com a wholly separate entity with its own president and a separate sales staff.

The dilemmas for AccuWeather were of a different nature. In Philadelphia and New York, KYW and WINS were paying good money for information that listeners could find on accuweather.com. Unless they particularly liked to listen to a forecaster, and both Elliot Abrams and Joe Bastardi had followings, why should they bother to listen to KYW or WINS for their weather? In effect, the website was, and still is, competing with the broadcast clients who are helping AccuWeather pay its bills. Yet, it couldn't very well let its competitors—especially weather.com—have the internet to themselves.

AccuWeather opted for a hybrid approach, a 21st-century solution with elements of the 1960s in that it was a variant of the original Joel Myers model with a dash of the old Columbia Record Club. Online consumers could get basic forecasts, maps, and radar for free, but if they wanted more—say, a more detailed forecast that included an hourly readout for the next several days, they could pay a premium fee of $7.95 a month. If they wanted the whole salami—access to the European model output, expert analysis—they could go all the way and buy the "professional" service for $24.95 a month. AccuWeather offered a free trial, and if users liked it and wanted to continue getting the service, they would have to pay, reminiscent of the deal Myers had offered ski operators in the great way back. It also included the Columbia Records–style negative option: If you failed to notify the company that you didn't want the next offering, you were going to get it and be responsible for paying for it. Similarly, the onus was on the subscriber to inform AccuWeather that the higher-grade service no longer was wanted.

One of AccuWeather's prime attractions was the Snow Man, himself, the voluble, prolific, and indefatigable Joe Bastardi, he who saw storms that not even a computer could see. Bastardi now has a national presence as an outspoken critic of man-made climate change, appearing on national television shows and having a running public debate with Bill Nye "the Science Guy." He has tangled with environmentalists and compared his bouts with them to his experiences during his wrestling days at Penn State. He says his adversaries would have been no match for the ones he encountered on the mats in college. He has been a pariah in the environmental community. His own company quite publicly distanced itself from his views, officially declaring that climate change and its assorted consequences were matters of open debate. But for many AccuWeather regulars, Joe Bastardi was the reason they visited the site, something that Joel Myers was well aware of.

Bastardi posted columns every day, sometimes 1,000 words, sometimes more, sometimes coherent, sometimes not, paralleling his breathless and entertaining style exhibited in his broadcast feeds. I spoke with him frequently throughout the years on the phone and in person, and whatever else one could say about Joe Bastardi, he is never dull. When I first met him at the AccuWeather offices, I would say he did not resemble the weathermen I had seen on TV, or Elliot Abrams, or Joel Myers, for the matter. He was dressed in shorts and a T-shirt, wearing a Yankees cap. He was the son of a meteorologist and had grown up in Providence, Rhode Island, and lived in Somers Point, near the Jersey Shore, in his youth, so he was well familiar with AccuWeather major markets and acquainted with the heartbreak of a snowstorm bust. In his broadcasts and columns, he often tried to affect a traditional adult distaste for snow, like the weathermen I had seen on TV. Joe Bastardi wasn't fooling anyone. He was as energized about the prospects of snow as he was about his weightlifting, which is to say he had the energy of a storm bombing off the coast. Bastardi was a competitive and passionate weightlifter, as his readers well knew. His forecast musings often were off the mark, but he would be ready with a concession speech, and when he busted his columns would include a "take out the trash" section.

He had some major hits. For the Presidents' Day Storm II of 2003, he was well ahead of about everyone in calling for 12 to 20 inches of snow in the East. Those rare snow events were, by definition, rare, but even in the barren times, Bastardi's "snow geese" could turn to him for hope. In mid-October 2009, Bastardi boldly predicted that the Mid-Atlantic region would be in for a cold and stormy winter. The region was crushed by record snows in December and February; the winter of 2009–2010 was the snowiest on record in Philadelphia, an outcome captured in AccuWeather's online traffic.

Bastardi's critics would argue that his scores were the result of the principle that a stopped clock would be right twice a day. He often has been accused of hype-mongering, but from my lay perspective he usually had scientific reasoning to back up his speculations. Also, I don't know of anyone who put in more time studying maps and data, and seeking out patterns in the chaos, in between pumping iron. His supervisor at AccuWeather, the late Ken Reeves, told me that one day he ordered Bastardi to just go home; the man was working on fumes.[35]

Seeing Bastardi's popularity, Myers made the smart business decision and planted his rainmaker (or snowmaker) behind the wall in the early 2000s. If the customers wanted a daily dose of Joe, they would have to pay the $25 a month "professional" fee.[36] Myers did not say how many customers paid that price specifically for Bastardi, but he did allow that the numbers were substantial.

And Myers admitted that when Bastardi abruptly quit AccuWeather in February 2011, after 32 years with the company, the business took a hit. His departure was a shock to his colleagues. Bastardi had idolized Myers; that was evident in his comments about him, and Joe Bastardi is no phony. He has never talked publicly about why he resigned. He left without so much as a goodbye to any of the forecasters he left behind.

Bastardi joined a startup company, Weatherbell, teaming up with Joe D'Aleo, he who had helped John Coleman develop the TWC concept. Coincidentally, he was an alumnus of Weather Services, Inc., which later became part of The Weather Company. As Bastardi, D'Aleo has raised questions about the sources of climate change. Weatherbell's online business doesn't remotely rival AccuWeather's or weather.com since it is strictly a paid service and never had an entitlement audience. It does publicly post Bastardi's winter and hurricane outlooks, and he is clearly the face of the company, making frequent TV appearances.

In a November 2018 interview on a Harrisburg TV station, it was evident that when he left AccuWeather, he took the snow with him. There he was in classic JB style saying that the prewinter "analogs," how the upper-air and sea-surface temperatures were aligning, were similar to that of the most brutal winters on record in the East. Furthermore, he said he saw "antilogs," the patterns that were the mirror opposite of those that preceded the tamest winters. By then he had dropped any pretense that he was rooting against winter mayhem. "You know a good winter for a snow-lover like me is lots of snow," he told the interviewer, "just keep bringing it. But that may be a bad winter to others."[37]

It would also be quite a good winter for The Weather Company, AccuWeather, and Weatherbell, but that wasn't the case in the winter of 2018–2019, in the I-95 population centers, where snowfall was below normal in Boston, New York, and Philadelphia, and only slightly above normal in the Washington area, home of the government's weather Pentagon.

Perhaps it is coincidental or has something to do with the East's winter-weather volatility, propensity for snow, and population densities, but the East appears to be a favored location for weather centers—including those of the U.S. government.

But it does, indeed, snow in the West, where snow arguably is more important than it is on the other side of the country.

CHAPTER EIGHT

Westward Snow

The Great White Reservoir

On the day James E. Church arrived in Reno, Nevada, he almost left. He no doubt already had misgivings about moving out West. He was a professor of Latin and classics, and had accepted a job at a university with a seven-year-old campus. Whatever misgivings he had were intensified on that first day. The professor witnessed a man dropping dead in the street, the victim of a saloon gunfight. It was like something right out of a movie, although this was reality; this was 1892, well before anyone figured out how to put pictures in motion. Church decided then and there he would head back to his native southeastern Michigan to rejoin his high school sweetheart. But on his way out of town something stopped him. He caught sight of Mount Rose in the majestic Sierra Nevada range. "He was so captivated by its beauty that he changed his mind and stayed," said the man who told that story to a reporter, James E. Church historian Tim Gorelangton.[1]

It would be impossible to know just how much of that tale is true and how much apocryphal, but beyond dispute, generations of Westerners have prospered from Church's decision to stay in Reno. As in the case of Wilson A. Bentley, James E. Church had no formal scientific training and could be self-effacing. Of his fascination with Mount Rose and his explorations he once made a Bentley-esque assessment, almost as though he were quoting the Snowflake Man directly: "To the public I was merely a great fool."[2] Maybe it's that inevitable association of snow with childhood—think of how snow and his sled, "Rosebud," evoked memories of childhood for John Foster Kane in the movie *Citizen Kane*—that might make an adult feel a certain foolishness for being so "captivated" by snow. As Bentley, Church was an unlikely candidate to advance the cause of snow science. Yet, his insights, determination, and assiduous research had a profound impact on the West as we know it today. Whatever misgivings he had upon his arrival, he clearly got over them. He would spend more than 40 years at the University of Nevada in Reno.

Church was well-acquainted with winter by the time he moved to Reno. He had lived in southeastern Michigan, about 50 miles northwest of Detroit, in the town of Holly. It wasn't in the extreme Great Lake–effect Snow Belt, but it experienced winter aplenty. The Upper Midwest has a tremendous variety of weather with those occasional punishing blizzards, prolonged snow covers, subzero temperatures, and assorted winter inconveniences. Those often-extreme conditions wrought by the interplay of jet-stream winds; the vast expanses of freshwater on lakes Superior, Michigan, and Huron; and the "Colorado Lows" and frontal cyclones that lure moisture-laden air from the Gulf of Mexico. Chicago is not only the "crossroads" of America for its crisscrossing rail lines. Carl Gustaf Rossby's high-powered meteorological shop at the University of Chicago happened to be situated beneath an atmospheric crossroads, a favored location for the weather-making polar jet stream.[3]

One thing the Upper Midwest does not have is mountains, and I can imagine what a stunning vista it must have been for James Church. I know I will never forget the first time I beheld mountains in the American West; "captivated" would be an appropriate word.

Church had earned his bachelor's degree from the University of Michigan in Ann Arbor, not far from his hometown, and landed a job teaching classics at the University of Nevada, 2,200 miles from his hometown. Appropriate to his last name, one of his specialties was Latin, which he taught, along with Greek and art history. Given his knowledge of Latin, he no doubt understood the Romance etymology of the names of the state and the mountains. Nevada is derived from the Latin root for *snow* and is Spanish for "snow-covered." In addition to his academic pursuits, Church developed an intense, passionate interest in the Sierras.

Church would come to know Mount Rose thoroughly. He made his first ascent on New Year's Eve 1895, with a colleague, a "trip made solely for the love of winter nature," he would recall. He hiked there frequently and published his experiences in the Sierra Club newsletter.[4] He became the first person of European descent to reach the 10,788-foot summit of Mount Rose, which overlooks Reno and Lake Tahoe. He and his wife spent Christmas week 1901 camping near the summit. He recalled how incredibly peaceful and serenely quiet it was. The only sounds they could hear were those of distant locomotives. They were the sounds of the future. The West was undergoing a population explosion, and Church would one day help the frontier accommodate its fresh arrivals.

It was on Mount Rose that James Church would influence the history of the West. As they had in the East, the railroads were transforming the once-unpopulated—and chronically parched—territories. It is not a stretch to say that Church was at the center of a version of the saga of *How the West Was Won*, a version far closer to reality than the six-gun legends.

In the East, snow can be disruption, enchantment, beauty. In the arid West, snow is white gold. The essence of snow is part of the essence of life in the West. Snow is water in magnificent disguise. The East is blessed with vast water resources, sometimes more than it can handle. Rarely is annual precipitation in the populous cities less than 15 percent of the long-term normal values. On average, for example, New York City, with its proximity to the Atlantic Ocean and hosting two major rivers, receives almost 50 inches of rain, as well as melted snow and ice, a year.[5] Las Vegas survives on a grand total of 2.4 inches of precipitation. New York typically gets more rain in three weeks than Las Vegas receives in 12 months.[6] When it is soaked by tropical-storm remnants, in a single day New York city might see twice as much rain as Las Vegas does in an entire year. Denver, to the east, is wetter, but not by that much; its annual totals are well less than a third of New York's.[7]

Where do the residents of the West and the millions of tourists get their water? It is all about snow. In the East, precipitation is not only plentiful, but also generously distributed throughout the year; on average, the Northeast can count on at least three inches a month. In much of the West, not only are the annual amounts paltry, but also spring and summer are drier than the cool seasons, with only about 10 percent of annual precipitation falling during the growing seasons. So, the residents, businesses, and utilities are prone to look up to the peaks. As one government water specialist in Idaho remarked, "The snowpack is like a giant reservoir in the mountains."[8]

Not surprisingly, water has long been a source of contention. Mark Twain, who worked briefly for the newspaper in Virginia City, Nevada, about 25 miles south of Reno, allegedly once remarked, "Whiskey is for drinking; water is for fighting over." Even if neither the attribution nor the precise wording can be verified, the quotation does more than hold water. The classic TV Western *Bonanza* was supposed to be set in Virginia City, and in its 400-plus episodes, viewers saw way more dust than rain.

In the early 20th century, when the West was undergoing its historic growth spurt, the tensions escalated to boiling points. One consequence was what became known as the "Lake Tahoe Water War." It involved a re-markably motley set of combatants that included farmers, a power company, property owners, and Native Americans.

Lake Tahoe, framed spectacularly by the Sierras, is a blue oasis in other-wise rain-starved Nevada, the second-deepest lake in the United States, at 1,645 feet.[9] It prospers from those frequent profoundly heavy snows of the Sierras. Mountains are precipitation hogs. In mountainous areas, it is not unusual for some of the world's more rain-deprived places to be situated in proximity to the wettest, as in the cases of Hawaii and Chile. Precipitation forms as the water vapor in air that has been warmed by solar energy at the surface, rises into cooler air, and condenses into rain and snow. Mountain slopes have an "orographic" effect, giving the rising air a lift and, in effect, forcing it to wring out its moisture at the summit and higher elevations on the windward side. This is most evident in winter when snow-capped moun-tains overlook dry valleys. From the air, the mountains appear to be rippled with white, those snow caps appearing as the crests of white waves washing over the peaks.

The orographic effect also erects a "rain shield" that deprives areas on the lea side of the mountains of snow and rain. The windward flank of the Pacific, which forms a barrier to winds off the Pacific, receives 10 times the precipitation of the areas to the east.[10]

The summit of New Hampshire's Mount Washington, in the Presi-dential Range of the White Mountains on the northern edge of the Ap-palachian chain, has a reputation for having the worst weather on Earth. It might well have the worst *observed* weather. Taking observations on a summit isn't like taking them at the airport. Church once set up an obser-vation station at Mount Rose, but it didn't last that long and didn't com-pare to Mount Washington's, which is a fully equipped weather station and research center that is staffed throughout the year, 24 hours a day. It does hold the record for the strongest verified sustained wind—235 miles per hour. In all likelihood, however, stronger winds have occurred on Rocky summits. Winds increase with height, as hikers well know. Mount Wash-ington stands at a little more than 6,200 feet. The Mount Rose summit is better than 4,500 feet taller. In winter, that summit would be a forbidding and frightening place for the average human being of sound mind who had a desire to live to see spring.

James E. Church was not the average human being. Among many other things, Church was a mountain hiker. That is a special breed of human; I know because we have two of them in our family, one of them an inveterate winter hiker. Personally, I've done enough mountain hiking to know that you would not find me anywhere near the Mount Rose summit anytime between the fall equinox and the summer solstice. Winters on mountains can be perilous. For verification I would recommend *Not Without Peril*, by Nicholas Howe, a collection of cautionary and, in some cases, frightening tales about the hazards of winter hiking in the White Mountains. Ascending and, especially, descending require that rarest of combinations: a meditative disposition and undivided attention. In spring and summer, a slip could cost you a broken bone; in winter, it could cost you your life. The snow amounts in the Sierras are prodigious—in the winter of 1906–1907, more than 850 inches,[11] or about 30 New York winters' worth, accumulated in the mountains—obviously at different depths at different locations, but that was the overall estimate. That might be as difficult for an Easterner to fathom as the concept of a major city and national tourist mecca somehow getting by on 2.4 inches of rain a year.

As it is in most of the United States outside of desert areas, nature can be generous or parsimonious with winter precipitation, and snow amounts in the Sierras can vary radically from year to year, as they do in the Eastern mountains. But that variability is a matter of tremendous significance in the West. A snow-challenged winter in the East might hurt the winter-sports business, but the residents in the valleys and the coastal plain are going to have plenty of water available almost every year.

About a third of California's water supply comes from Sierra snowmelt, according to estimates by the University of Southern California.[12] In the West, as much as 80 percent of precipitation falls between November and March, and more than 75 percent of Idaho's annual stream flow has its source in liquified mountain snowpack.[13]

The snows of the Sierras are a crucial part of the lifeblood of the West. So that *how much* question is of vital importance throughout the course of a season. In any given winter, for forecasters that question remains unanswerable. For some winters, the government's seasonal forecasts, which confine themselves to the limits of the science, can at least say that the odds favor above- or below-normal precipitation in the Western mountain ranges. In the 19th and 20th centuries, when the West was undergoing historic population growth, not even those ambiguous outlooks were available. For that matter, how many of the new arrivals paid much attention to the snow in the mountains, beyond its role as obstacle, or understood its importance?

James E. Church was one who did, and he devoted a significant portion of his life to answering the *how much* question in a very different and vital way.

Inches, Feet, and Reality

Understandably, when snow is in a forecast, the public tends to be fixated on inches and feet. They are the numbers that matter for the winter-sports industry and for plow operators and the thousands of school administrators who have to break the news to parents that they better find day care or they'll be stuck with the kids all day. Logically or not, accumulation forecasts affect the perception of a storm's severity, and we've seen again and again that a forecast can have a greater impact than the actual weather. Just ask restaurant owners and the people who run supermarkets. Accumulation estimates will always be a challenge, but I've discovered that strangely, the *how much* question can be difficult to resolve long after the snow *stops*. I would argue that those inches and feet projections receive far more attention than they merit. That clearly was something that James E. Church grasped intuitively.

A word about snow measurements: To the chagrin of climatologists like Colorado's Nolan Doesken, former head of the American Association of State Climatologists, and Rutgers University's David A. Robinson, an international expert on snow and ice cover and the nation's longest-serving state climatologist, measurements often are far more art than science. The National Oceanic and Atmospheric Administration (NOAA) has 14 pages of complex guidelines on how snow is to be measured, how frequently, and where. Methods have changed throughout the years, and the latest version was published in 2012. But it is unclear how often those standards are followed by observers who report totals to National Weather Service offices—those amounts to the tenth of an inch, say 8.7 inches, notwithstanding. Using the decimal figure gives a false sense of precision, like the car salesman who quotes an exact figure, say, $29,898.17. And the official snow reports from the government's first-order measuring stations, which theoretically are monitored for quality and corrected for errors, sometimes become sources of serious contention and controversy.

Take Philadelphia's standing record of 30.7 inches on January 7–8, 1996. The total was shocking, given that the old record was 21.3 inches and that Philadelphia International Airport's totals usually were lower than those of other spotter reports, partly the result of environmental and microclimate factors. After it was posted, observers throughout the region reported higher totals, including one of three feet across the river from Philly in South Jersey. Chet Henricksen, who was then in charge of the local National Weather Service office, was suspicious of those reports, convinced that at least some of the observers were playing a game of snow one-inch-manship.[14] Snow amounts are an inordinate and not necessarily explainable source of pride among snow-philes. He also had a solid basis for being skeptical of the airport totals: The snow was never actually measured.

I had visited the official snow-measuring station two winters earlier. It was an existential experience. It was located on the roof of a terminal building, which was next to a river and near a swamp. The roof of a heated building is not the most desirable of venues for measuring snow, and you won't find it recommended in NOAA's standard procedures. The observers couldn't very well measure snow on the nearby runways. ("You didn't want them to get run over by a damn plane," said Henricksen.)[15]

To my amazement, their solution in some cases was not to measure the snow at all. Instead, snow and sleet were captured in a rain gauge, equipped with anti-freeze to melt any frozen precipitation! The accumulation was estimated based on the amount of melted liquid and the air temperature. The conversion numbers were listed on a NOAA table. For example, an inch of rain at a temperature near freezing would translate to 10 inches of snow, although research has cast doubt on that assumption.

For that 1996 storm, during which temperatures remained in the teens much of the time, officially 1.55 inches of melted precipitation was measured at the airport. That rather tidily converted to 30.7 inches using the conversion chart. The National Weather Service eventually commissioned Robinson, the New Jersey climatologist, and Jon Nese, a former Franklin Institute meteorologist who by then had moved on to become a professor at Penn State, to investigate the total. Upon further review, they did not find sufficient evidence to overturn the call. So the call stood, and it remains on the books.[16]

The NOAA table may well have some value; however, it relies only on temperatures in the lower levels of the atmosphere. Snow accumulations are heavily dependent on the types of flakes, which are affected by the layers of atmosphere through which a crystal travels in a journey that can take as long as three hours—not just by conditions near the surface.

Accumulations also are greatly dependent on the water content of the crystals, and that is the crucial component in the West. A seasonal total of 800-plus inches may be attention-getting, but in the West that is not what hydrologists, farmers, and various public officials needed to know. What matters is precisely how much liquid is in all that snow on those gleaming majestic slopes and summits. It is a complicated question.

James Church was not seeking to address this grand enigma when he first climbed Mount Rose in snowshoes in 1895. What he sought was one of the most spectacular vistas in North America. Church was enraptured by the Sierras, which he once described as a "wonderland so remote from the ordinary experiences of life that the traveler unconsciously deemed that he was entering another world."[17]

Church left Nevada for an important interval, to marry that high school sweetheart, in Michigan, and pursue graduate studies. He eventually earned his doctorate in Munich, where he and his wife had ready access to the Alps, where they hiked. His doctoral dissertation would have nothing to do with snow or the mountains. It was a treatise on Latin epitaphs, not the stuff of ski lodge chatter over a snifter of cognac. He returned to Nevada as Dr. Church, often hiking in that Sierra "wonderland"—and stepping into a very worldly thicket in the new century.

Lake Tahoe had become a prime resource for irrigation, hydroelectric power, fishing waters for Native American Paiutes, and unwelcome floodwaters for property owners. The lake was the source of the Truckee River, which flowed for 70 miles before emptying into Pyramid Lake to the northeast. Near the end of the Civil War, a dam was built at the Truckee headwaters, and the so-called Newlands Reclamation Act of 1902 provided funding for the construction of additional dams for irrigation for the otherwise rainshadowed arid lands to the east of Reno. Thus, farmers had a significant stake in the career of Lake Tahoe, as did utility customers, as did property owners, as did the Native Americans, who had a guarantee that the government would provide sufficient water for their fishing needs. Those crucial lake levels were dependent on the volatile and unpredictable volumes of meltwaters from the annual snowpack. *How much for the Sierras* wasn't about snow so much as it was about water.[18]

Humanist as Scientist

Hikers have to be acutely aware of the weather. Elevation brings with it a certain light-headedness, which can be a pleasant experience, even exhilarating. But a mountain hike also can be a precarious adventure coursing through four seasons. Along with being a passionate hiker, Church was intensely attuned to his environment and had a well-cultivated interest in meteorology. His interest was piqued by an article he read by a meteorologist at the U.S. Weather Bureau's San Francisco office, Alexander McAdie, who called for setting up weather stations in the highest peaks of the Sierras.[19]

With a modest federal grant and thermometers supplied by the federal government, Church and a group of associates set about constructing an observation station atop Mount Rose, a more realistic possibility than erecting one on Mount Whitney or the summits of other higher peaks. In 1906, he volunteered to climb Mount Rose every month for a year to obtain temperature readings. "The study of snow was begun," he would recall later.[20]

He presented his findings in an article in the June 1906 issue of the *Monthly Weather Review*, one of the country's most important scientific journals. Evocative of the modesty of Wilson A. "Snowflake" Bentley, Church described himself as an "amateur" who had been invited to submit his paper by the editor, Cleveland Abbe, a towering figure in the history of meteorology. Hauling the equipment and erecting a shelter at the summit was an ordeal, but using a maximum–minimum thermometer, Church was able to begin recording temperatures in the summer of 1905. He took readings more or less monthly; the low for the period from August 4 to September 4 was "so startling" that, according to Church, "I immediately began to search for some defect in the instrument or the shelter." He thought that perhaps summit gales had knocked it out of calibration; however, he concluded that if anything, that might have made the thermometer read high. The low for the period from September 6 to October 7 was 4.5 below Fahrenheit. Church viewed that as "authenticated" by a reading of 1 below atop a Sierra summit that was about 2,000 feet lower.

Church was just getting started. He was about to undertake a project that would lead to one of the most significant developments in the history of the West.[21]

The relationship between snowpack and meltwater was an obvious one. Not so obvious was the relationship between the depth of the snowpack and the amount of water contained therein. Not only do snow crystals vary in shape and size, as Wilson Bentley had documented with photographic images, but also they can vary tremendously in the amount of water they hold, depending on the atmospheric conditions they have experienced on their earthward journeys. The wetter the snow, the heavier it is. Experienced shovelers can tell you that a four-inch snow in March, when temperatures are borderline, can be a far greater challenge than eight inches of snow in January, when temperatures are in the teens and low and mid-20s. Plow contractors typically charge by accumulations; it might make more sense if they charged by the water content of the snow, because that speaks to the weight that the plows have to move.

As Church deduced, the weight would speak to the water content. An inch of water on a square-foot surface weighs 5.2 pounds. It follows that any depth of snow atop a square-foot surface that weighs 5.2 pounds would contain an inch of water. It didn't matter whether it was six inches of soggy, wet, March-type snow or 20 inches of January powder with temperatures in the teens. The snow was the show; the water was the reality.

Church first got into this measuring business to settle an argument involving loggers and pioneering conservationist Gifford Pinchot, who was the nation's chief forester and later went on to become governor of Pennsylvania. (He also was a prohibitionist whose legacy included some of the nation's strictest and strangest liquor laws.) The loggers held that they were contributing to water supplies by cutting down conifers. The trees, they argued, were giant straws sucking moisture out of the melting snow. Church disproved that hypothesis, demonstrating that the conifers had no impact on the snowpack.[22]

So how did Church do it, and how did he figure out how to estimate the water content of the trillions and trillions of tons of snow that landed on the Sierras? And how did he communicate a figure that would have meaning to the people who were relying on all that water for consumption, agriculture, and power? How meaningful would it be to say, "Well, folks, this year we can expect about 10^{15} tons of melt water. Enjoy."

The obstacles to the basic measurement of snow near the summits would be intimidating, starting with the wind. The summits and higher elevations are constantly subject to hurricane- and tornadic-force gusts. Given the frigid nature of the temperatures, the snow would tend to be powdery and thus given to chaotic flight, and the snow-to-liquid ratios would be high. The NOAA instructions for measuring snow are explicit in directing observers to take their readings in clear areas, free of trees and other obstructions. That was not going to work in the Sierras. Church discovered that in the case of the Sierra environment, the forested slopes, where snow would be less susceptible to high winds, were his best options for taking measurements. Once snow lands, it undergoes complicated changes. It compacts and settles, and those crystal masterpieces assume a homelier, prosaic, and denser identity. They might get rained upon, and I learned from the masters that rain doesn't wash away snow—melting is a more complicated and elegant process—but can add considerably to a snowpack's water content.

Generally, the snowpack would be several feet deep, so rulers and yardsticks weren't going to cut it. Church set about to invent something that would. His solution was a metal tube device, as many as 10 feet long, that resembled a flute, with cutting teeth etched into the bottom to drive into the snowpack. Additional tubes could be attached with screws. A wrench device could drive it through the frozen mass until the teeth met the soil. The tube, about 1.5 inches in diameter on the inside, was slotted, so he could see what was inside. It was marked with inch markers so that he would easily know the depth. Once it was packed with snow, Church would weigh the tube on a specially calibrated scale, subtracting the weight of the tube.

The diameter of the tube was important: An ounce of snow at that width would convert to about an inch of water. The readings would yield the ratio of snow to liquid in a given location. Perhaps a 12-inch "tube" of snow would weigh an ounce, yielding that inch, or it might be a six-inch tube of snow. Again, it was about the weight, not the snow depth.[23] By setting up snow "courses," say, 5 to 10 of those measuring devices for a distance of 1,000 feet, he could calculate the liquid content of that vast expanse of snow.

Church was confident enough in the accuracy of his calculations that in February 1909, he announced that he had invented a device that he named the Mount Rose Sampler. Mergen described the name as a master stroke. "While other gauges bore the names of their inventors or were simply called snow density apparatus, Church evoked some of the romance of winter climb and called attention to the uniqueness of the place where snow surveying was born."[24]

Three other men had developed similar devices at about the same time, but as Mergen has said, Church had the rightful claim as the first. Church persisted in perfecting his system, and in staking his claim as being the origi- nator, it helped that he was a prolific and excellent writer with a flair for the poetical. In one of his papers he referenced the "abundance of knowledge that nature will pour into your lap if you will come within her reach."[25] All that classics training must have come in handy.

Church benefited from the counsel of the inventors of a precipitation gauge used by the U.S. Weather Bureau. The resources of the university were immensely helpful. The stainless steel tubing he used for his apparatus was assembled in the machine shop of the university's engineering department. To work up the intricate calculations to convert his measurements into projections for water-runoff potential, he consulted with a civil engineering professor.[26]

Church's breakthrough invention turned out to be invaluable. Efforts to measure the meltwater content of the snowpack dated to the 1830s, but Church's sampler inaugurated an era of systematic water surveying in the West. As it turned out, the spring water levels of Lake Tahoe correlated nicely with the water equivalent of the snowpack in late March, and that was evident in 1911.

The Sierra snowpack surpassed 37 feet in some areas in March of that year. A back-of-the-envelope calculation for the water content of the snow would have been dangerously far off the mark. In February, heavy rains fell atop snow, which absorbed it like a sponge. The idea that rain washes away snow is largely a myth, as Nolan Doesken explained to me. In addition to being a national authority on snow measurement (and, yes, he says, most people are doing it wrongly), he is an expert on the all-important melting process.

It's Not the Heat, It's the Dew Point

An Easterner skiing in the Rockies might marvel at how the snow can withstand the unfiltered sun directing its energy on the slope on a clear day when the temperature climbs to 50. If the air is dry enough, the dew point is low, and snow has a hard time melting. The dew point—that's the temperature at which water comes out of hiding and condenses, as it does on dewy mornings in the fall when water droplets form on the foliage and car roofs—is the catalyst for melting. The relative humidity, which by definition is "relative" to the amount of vapor that the air can hold at a certain temperature, just confuses things. For melting purposes, the key is to know at which point water vapor will work its magic and change state. When water evaporates, it gives off a cooling effect; when it condenses, the opposite occurs. The latent heat stored when vapor becomes droplets releases heat. When the air temperature is 50 and the dew point is low, the melting is slow. Ski areas can make snow even when surface temperatures are several degrees above freezing if the air is dry enough; the air cools as some of the moisture emitted from a machine evaporates. When the dew point is 50, vapor condenses in a hurry when it comes in contact with the cool surface of the snow, and the resulting heat is an efficient melting agent. The air becomes a blowtorch on the snowpack. In the Northeast, this effect was particularly evident after the January 1996 snowstorm. In advance of a rainstorm, the dew points shot up dramatically, and the immense snow cover all but disappeared in a matter of hours. And this was before the rains even started. The result was widespread destructive riverine flooding. What mattered to the height of the river crests was the amount of liquid in the snowpack. The rains did not wash away the snow; it was gone before the rains came.[27]

In 1911, the February rains in the Sierras didn't erase the snowpack; they just made it all the soggier. Church was well aware of the impending crisis thanks to his survey methods. His warnings to the Truckee River General Electric Co., the keepers of the Tahoe dam, prevented a potentially disastrous flood. And with Church's surveys in hand, dam operators were able to regulate the releases of flood waters.

Throughout time, the net result was a *pax aqua* as the predictability dampened tensions among the residents along the lake, the power company, and Native Americans. Church's methods became a prototype, and during the next 30 years, a surveying network blossomed throughout the West.[28]

Church became a widower in 1922, and he became an international authority and consultant on snow research, traveling to Russia, Northern Europe, and the Himalayas.

"I had gone to the hills for pictures and pleasure," Church wrote, and he ended up "giving birth" to a worldwide organization, the International Commission of Snow.[29] He was president of the commission, but because he had become ill while in Russia he was unable to attend the first meeting in Edinburgh in September 1937. He did submit a paper, "Snow Surveying: Its Methods and Principles." It described the "economic importance of the snowfall in the high mountain ranges of the western United States to the semiarid plains below."[30]

The government continues to use a variant of the Mount Rose Water Sampler to estimate the meltwater potential. It employs a vast network of devices that use essentially the same principles, these days supplemented by satellite and other high-tech data. In 2006, the Department of Agriculture presented a plaque to the University of Nevada honoring Church's accomplishments. "This information is crucial to everyone who uses or drinks water," declared the deputy agriculture undersecretary, Merlyn Carlson, under President George W. Bush. And Carlson said that generations of Westerners should be grateful to the Latin professor who fell in love with the Sierras. "Thanks to him, we have 100 years of snow-surveying data."[31]

Yes, It Snows Below Also

> The air was so thick with fine-ground wind-lashed ice crystals that people could not breathe. The ice dust webbed their eyelashes and sealed their eyes shut.—David Laskin, *The Children's Blizzard*[32]

The Rocky ranges that tower over the lowlands from the U.S. Southwest to western Canada indeed are the great monopolizers of Pacific moisture, depriving their eastern flanks and valleys of precipitation; however, two notorious cyclonic perennials, energized by the jet-stream winds rushing out of those frigid regions of the Arctic and Yukon, historically have conspired to make winters punishing and adventurous throughout the vast expanses that we commonly call the Great Plains, that massive bowl of low country between the eastern and western mountain chains. Easterners are well familiar with the "Alberta Clippers" that originate in the Pacific and dive into the Plains states from the Canadian border and race toward the East Coast. Along the way, they are prone to generate a corridor of nuisance snows, with the Appalachians acting as a poor man's Rockies, intercepting whatever residual moisture they have retained. Sometimes, however, a clipper can transfer its energy and spawn a secondary low off the Atlantic Coast that can transform into a major nor'easter, a storm known as a "Miller B," so named for a researcher who developed the classification system.

Easterners likely would be unfamiliar with another instigator of disruption, the "Colorado Low," with which Westerners and Midwesterners would be well acquainted. Those cyclones can develop explosively, creating a reign of terror of heavy snows and frigid north winds. They can affect tens of thousands of square miles from Denver to Chicago. When the circulation around the center expands, the impacts can penetrate all the way into Texas, and they take on another identity as "Blue Northers."

One of the most notorious Colorado Lows in the period of record had a direct impact on Ken Libbrecht and the other residents of Fargo, North Dakota. It approached South Dakota on the morning of March 3, 1966. The heaviest snow and strongest winds usually occur to the north and west of the storm center, and that was true in this case. Snow had developed the day before in southern North Dakota and crept northward. Blinding curtains of snow paralyzed Fargo. Technically a blizzard requires three hours of visibilities of a quarter mile or less. Incredibly, driven by winds as fast as 50 miles per hour, Fargo endured 30 consecutive hours of blizzard conditions from March 3–5.[33]

The gale-driven snows piled to near the top of utility poles in Fargo. That is not hyperbole; the image has been frozen in a famous photographic image. Grand Forks measured almost 28 inches, which remains a record for a single snowstorm. Temperatures were in the 20s when the snow started and sank into the teens at the height of the storm. Yet, the National Weather Service poststorm account described those temperatures as "mild."[34] Those north winds to the west of Colorado Lows can import Arctic air that Easterners rarely experience during snowstorms.

The Rockies form a barrier to winds from the west off the Pacific, and in the East, the Appalachians have a moderating effect on the coldest Canadian air masses before they reach the population corridors. But in the Plains, nothing blocks a bitter wind from the north, the snow and ice cover over the lowlands acting as a superhighway for channeling polar invasions. Temperatures can fall below zero during snowstorms, and conditions during the 1966 storm constituted a day at the beach compared with the blizzard of January 9, 1887, when temperatures plummeted to as low as 50 below. The cold that winter was so horrific that it resulted in what became known as the "Great Dieup," a ghoulish play on the term the "Great Roundup." More than

350,000 cattle perished in that winter in Montana alone.[35] As severe as that January storm was, something far worse would occur a year later.

The "Blizzard of 1888" is a term that instantly evokes New York and the Northeast. For its deadliness, impacts, and long-term consequences, that nor'easter justifiably holds an elite position in the winter hall of fame. But it may not have been the deadliest storm in the nation that winter.

Two months earlier, a blizzard ignited by a Colorado Low blitzed the Plains states from Nebraska to the Dakotas to Minnesota. The prairie lands were far-less populated and settled than the urbanized Northeast, which made the January blizzard all the more isolating and terrorizing. As David Laskin wrote in *The Children's Blizzard*, a riveting and deeply reported account of the storm and the historical context in which it occurred, the prairies were becoming populated with new arrivals. "Railroad companies flushed with millions of acres of government land grants promised new settlers the sky and sold them the earth at irresistible prices. . . . The settlers . . . banked on the future and put their trust in land they loved but didn't really understand." The newcomers lacked familiarity with the local "folk wisdom" that might attune a native to impending dangers.[36]

Unlike the March blizzard in the East, this was not a long-duration event. The snow lasted maybe 12 to 18 hours; however, at its peak it was more ferocious than its March counterpart. Whether it was deadlier would be impossible to know for certain. By all accounts it was even more of a surprise attack than the Eastern storm, and that element unquestionably added tragically to the fatalities. It became known as the "Children's Blizzard" or the "Schoolchildren's Blizzard," and today such a label might be interpreted to mean a magical storm that closed all the schools. But it is a deceptively benign name for a killer storm that claimed untold numbers of children among its hundreds of victims.

Normally, January 12 would be in the heart of the coldest time of year in the Plains, a copious importer of Arctic air masses gliding north to south along freeways of snow-and-ice-covered land masses. But the morning of January 12, 1888, was an extraordinarily mild one when a young teacher named Minnie Freeman arrived at her school in the town of Ord, Nebraska, about 150 miles northwest of Lincoln. Nothing foretold the horror that was to come.

To the north, the storm invaded the Black Hills around 10 a.m. "It began on warm morning," reads the fragmented note to the Weather Bureau from the late Frank Thomson, who lived in Spearfish, South Dakota. "Storm slid under the warm air." You could plow through a lot of meteorology books and not encounter a tidier summary of what actually happened: The heavy, dense, frigid, and out-of-control polar mass spilling southward forced the warmer air at the surface to rise rapidly, detonating a rapidly intensifying cyclone. "Snow like flour," Thomson wrote, "could not breathe in it. I was seven years and stuck my head around corner of house and nearly choked before I got indoors again."[37]

To the south in neighboring Nebraska, conditions deteriorated with a tornadic speed at about lunchtime. One Nebraskan said the snow was so blinding, "It was like trying to see with my face pushed into a snowdrift." The storm had arrived with "astonishing quickness." Said one survivor, "It seemed as though a wall of snow came down from the northwest." Said another, "The sun went out like an extinguished lamp."[38]

A man inside a school in Colfax County remembered the storm announcing itself with the "suddenness of a clap of thunder." He added, "The blizzard crashed against the schoolhouse like a tidal wave, shaking the wooden frame building and almost lifting it from the foundation."[39] The noise was terrifying, disorienting. "One could hardly make himself heard above the roar of the wind," said a Holt County man.

Among the victims were three pupils who attended a Plainview school. A teacher tried to lead them to safety in a boarding house 200 yards away. Tragically, they lost their way, and the children froze to death. The teacher somehow managed to survive the ordeal—and live with the nightmarish memories.

"Many a Nebraska home was anxious that night because of schoolchildren who were not in their own beds," recalled a Butler County survivor.[40]

A Teacher Stronger Than the Storm

Fifteen children who did live through the storm attended that school in Ord where Minnie Freeman taught. She is credited with saving their lives—credited too much, according to her—and in the process she became a legendary figure as her fame spread throughout the country. Her heroism was memorialized in a popular song. Her reputation has received a fresh boost in the 21st century. In 2006, the *New York Times* included her saga in its "Overlooked" series of obituaries, which paid tribute to forgotten female historical figures.[41] Her exploits were further chronicled in David Laskin's book.

I've written enough obituaries in my career to know that her actions during the blizzard would make for an irresistible story, and the *Times* obviously felt similarly. (I like to say that great minds think alike, but the reality is that mediocre minds are more likely to think alike.)

In Robert Frost's laconic "Fire and Ice," the narrator muses about the question of what would be worse, to perish by fire or ice. His poem came to mind when I read the chilling accounts of that blizzard. The crisis at the Ord school on January 12, 1888, had all the elements of a fire emergency, only in place of flames, the lives of the occupants were endangered by stinging snow crystals and polar cold. The winds, maybe 60 miles per hour, with gusts as fast as 80 miles per hour, ripped off a section of the schoolhouse roof and blew in the door, transforming the interior into a life-threatening ice box, the wind chills plummeting rapidly toward 40 below zero. If you've ever experienced an extreme wind you know just how disorienting it can be. To deafening winds, imagine adding the screams of a roomful of panicked children, ages 5 to 15, who just watched a part of the roof of their one-story schoolhouse go flying toward Texas. Minnie Freeman herself was a teenager, not much older than some of the pupils.

After the gales blew in the door, Minnie and a few of the boys in the class managed to get it back in place, only to have it blow in again. This time she had them nail it shut. What she did next is in some dispute. A famous story oft repeated is that among the school supplies she found rope, which she looped around each pupil and herself. Carrying the smallest child, she led them on a half-mile trek to the nearest farmhouse; however, according to Laskin, some other witnesses take issue with that account. One student wrote that Minnie had the students climb out a window and that they could have walked to a house a quarter-mile away with the winds at their backs. Instead, the group faced the wind and walked a half-mile to the boarding house where she was staying. "We were not tied together in any way, as has been erroneously stated so many times," the pupil said.[42]

Whatever action she took, she saved the children, and the children stayed together, and alive. Laskin pointed out that some of the older teachers failed to act as smartly or heroically. The details notwithstanding, what Minnie did that day qualifies as heroism.

"I've never felt such a wind," she said in an interview with a local newspaper. "It blew the snow so hard that the flakes stung your face like arrows."

Minnie Freeman became a national celebrity. According to the *Times* story, she allegedly received almost 200 marriage proposals and was inundated with gifts. She also was honored with what became a popular song: "Thirteen Were Saved; or Nebraska's Fearless Maid." And, yes, it bought into the rope narrative. Among the lyrics are the following:

> The brave girl gathered them about and prayed to God for aid,
> Then quick as thought from simple cord, a band of union made . . .

Then forth into the blinding storm, she led them bravely out,
One carried in her gentle arms, all cheered by word and shout.

She also is honored on the ceiling of the Nebraska State Capitol with an abstract mosaic representation of her rescue mission. For her part, she insisted that she was undeserving of the attention.

In a letter she wrote to the *Omaha Daily Bee*, she said, "Too much has already been said of an act of simple duty."[43]

CHAPTER NINE

Making It

The Art of Man-Made Snow

While the mountains of the West assume the crucial role of vast white reservoir, the mountains of the East have wholly different relationships with the lowlanders. In the East, save for a major drought about once every generation, water is almost everywhere. The Eastern mountain snows do replenish lakes and streams, but rain is so reliable in the populated coastal plain that stingy mountain snow seasons, even a sequence of them, won't precipitate a crisis. In the last decade, rainfall has been particularly robust, particularly in the Northeast,[1] and that could well be related to worldwide warming. More warmth would mean more water vapor available for precipitation. My various meteorological mentors throughout the years have convinced me to be wary of linear relationships, and, yes, the West has been experiencing tenacious droughts, but the warmth-water vapor relationships certainly make sense intuitively.

The Appalachian chain is essential to the East as a climatic buffer zone, for winter sports, for the tourism economies. But it is beyond fortunate that Easterners don't have to count on those mountains to be their reservoirs. Snow amounts in the higher elevations are wildly variable from year to year, and that is especially true in areas south of the White Mountains and the northern tiers of the Green Mountains and Adirondacks. Oftentimes has it occurred that the Poconos, in Pennsylvania, have gone begging for snow.

Yet, in some Appalachian winters the mountains are creamed with so much snow that skiers and hikers have to beware of "spruce traps," when accumulations are so profound that they cover evergreens. The areas above those buried treetops act like trap doors that can pull down the unsuspecting and confine them to a snow pile that can endanger their lives. In other seasons, the slopes in the East are hauntingly bare from October to April.

Even more so than in the Rockies, the weather in the Appalachians, being subject to several different storm tracks and so much available moisture, is subject to extreme seasonal and short-term volatility—sometimes very short-term. Ski veterans will tell you that when they head out in the morning for a day on the slopes, the pleasant conditions are the exception, and they learn to accept the realities of ice and plain, old wet rain on the slopes. They will tell you what a pleasure it can be skiing in the often brilliantly sunny and dry Rockies, when the snows are reluctant to melt even when afternoon temperatures soar into the 40s. The air in the East is frequently swollen with water vapor that condenses on the snowpack, resulting in the dreaded melting–refreezing cycle that can transform a slope into a high-speed downhill ice rink. In the West, that process usually is not a factor because the air is so dry and the dew points desert-like low.[2]

Ski operators in the Rocky and Appalachian chains do share at least one issue in that they are contending with a potentially game-changing variable. Compared with the 20th century, temperatures have risen 2 to 3 degrees Fahrenheit in North America, well outpacing the rates of rise more globally, and that is having an impact on ski seasons.[3] Interestingly, winter-snow extent throughout the Northern Hemisphere hasn't changed much in the last 30 years, according to climatologist David A. Robinson, who is an international authority on snow cover, and actually has increased in October and November. Snows, however, are vanishing earlier in the spring.[4]

Global warming is a likely suspect in almost any anomalous behavior by the atmosphere these days, but mild, snow-deprived winters are not a recent development in eastern North America, even in the Greens and Whites, and all the way up to Quebec and Ontario. While Easterners don't need that snow for their warm-season water supplies, the winter-sports industry sure needs snow for its cash-register supplies. That industry is important to regional economies and the U.S. Treasury, a source of federal, state, and local tax revenue that helps to pave roads and pay for schools in areas that otherwise might be deprived of ratables or other taxing resources. It also has collateral benefits for the likes of restaurants, gas stations, and gift shops. In the Poconos, it certainly is a boon to the billboard business. (We are grateful to Vermont for outlawing billboards, and we don't fully appreciate how much they blight the landscape until we cross the New York border and see those road signs once again detracting from the landscape.)

Holidays on Ice

December had been virtually snowless on Blue Mountain, a ski area in the southern vestibule of the Poconos. The frustration continued through the first week in January, when it was so warm the resort couldn't make snow. Bob Taylor, the commander in chief of the Blue Mountain snow guns, had seen this movie before, and it was among his least favorites. Finally, on January 9, temperatures dropped into the teens in the wake of a cold front and didn't get past the mid-30s for the next two days. Taylor's air cannons launched an all-out assault on the hills, spewing millions of snow crystals in the foreground of a crisp blue sky even as skiers and boarders crowded the hills on the first cold day in weeks. At long last, Blue Mountain was able to make snow. "Yeah," said Taylor, "for two whole days."[5] Talk about making hay when the sun shines. On the second weekend in January 2020, temperatures soared into the 60s. Yet, Blue Mountain was able to keep its trails open.

Even when the Eastern mountains are snubbed by nature, they do get their snow. Ski-resort operators have become very good at making their own. Throughout the nation, 90 percent of all ski areas rely to some degree on artificial snow, although far less so in the Rockies, where snow tends to have more staying power.

How inventors, ski operators, and entrepreneurs got so good at it is a labyrinthine tale replete with requisite gaps in the timelines and stories that might even be true. Based on the available evidence, snowmaking owes measures of gratitude to such diverse elements as the Dust Bowl–era droughts, the aviation industry, and the Florida orange groves.

Jeff Leich runs the New England Ski Museum in one of my favorite towns on the planet, Franconia, New Hampshire, on the west side of the Presidential Range. We go to Franconia as often as we can. That's the town where Robert Frost wrote some of his most memorable poems—notably "Stopping by the Woods on a Snowy Evening." And if you ever have the pleasure of walking or driving along the dirt road that leads to his former residence and the garden path marked with quotes from his poems, you'll see why he insisted that the previous owner vacate the premises so that he could claim the property for himself.[6] The only mystery would be why the previous owner agreed to that demand. The ski museum is at the foot of Cannon Mountain. We've hiked it and discovered that taking the tram is a whole lot easier way to get to the summit. Cannon draws the raw material for its snowmaking from the waters of magnificent Echo Lake, a popular swimming spot during the rather brief White Mountain summer, when you might see a bear and her cubs foraging for food along the verdant, shaved ski trail. No, it does not take its name from the snowmaking cannons, but from the peculiar shape of the summits.

We will not begrudge Leich his venue because he has made a heroic attempt to stitch together a snowmaking chronology from disparate sources and was generous with sharing materials. He dates a first attempt to 1934, when Dust Bowl droughts parched the mid-continent, although the first attempts weren't quite snow as we popularly know it, and the effort didn't occur in the United States, but in a country well acquainted with the vicissitudes of winter: Canada.

The Toronto Ski Club had planned a major ski-jumping competition that winter, but it encountered a crisis that would become a common issue for the winter-sports industry. Not only did the venue lack snow, but also none was in the forecast for the foreseeable future. Rather than surrender to nature, however, the club's leaders came up with a novel concept. The University of Toronto had a skating rink, and the university agreed to have its ice planer shave off layers of ice. Trucks delivered the shavings—75 tons of them—to the jump site, which was about four miles away.[7] Enough ice was available to cover the hill, with several inches available for the landing area. The run was faster than it would have been with natural snow, and the solution was economical, costing about $1 a ton; the university provided the ice for free. It is uncertain in whose mind the idea first crystallized, but at about the same time a similar project was undertaken on Bear Mountain in New York.

National Skiing magazine declared, "The real credit for the idea of improving on weather conditions belongs to the ski jumpers who began holding meets on crushed ice at Bear Mountain." Using ice for ski jumps became an "institution" on the mountain, and "seldom is a meet canceled because of the weather."[8]

The concept found its way to the coastal plain. Some of the most well-known figures in the sport came to Boston in November 1935, to ski—indoors. They participated in Boston Garden's first Winter Sports Show. For the event, 500 tons of snow were spread across a 17,000-square-foot course, 85 feet wide by 200 feet long. The show was such a hit that Boston followed with annual sequels. Not to be outdone, New York City hosted a winter sports show in Madison Square Garden in December 1936, which was so popular that "thousands" had to be turned away, according to a newspaper report.[9]

The ice did a more than serviceable job of mimicking snow, which isn't surprising; ice also is crystalline. Understandably, however, ski operators were in the market for something less labor intensive and more affordable. They would find it; it would just take a while.

One hint at a solution evidently was provided by the burgeonin, industry, so influential in both the history of meteorology and ti eration of the observation network. It serendipitously became inv .~u in "snowmaking" in the 1940s. Americans, Canadians, and Britons all were developing deicing equipment on their aircraft. That required fashioning nozzles that could produce the frozen materials needed for testing the effectiveness of ice-melters. While the nozzles weren't always spewing snow per se, it was a reasonable facsimile.[10] At the time, they evidently saw no commercial value in their created nuisance. Accounts differ on how much their work influenced snowmaking in the lower 48, but it is known that they did publish accounts of their adventures that drew interest in the United States.

The postwar period was pivotal for the future of human efforts to create snow. Americans were enjoying an unprecedented economic prosperity. Windows of opportunity were opening to take advantage of "disposable" incomes. The ski industry was poised to prosper, but it had no control of the weather. If money couldn't buy happiness, could it at least buy snow? Some believed the answer was yes. And given that snow was at times as unpredictable as a stock price, the pursuit of weatherproofing the ski industry intensified.

For a second consecutive season, the winter of 1949–1950 was maddeningly snowless in the Northeast, a condition as debilitating to the ski industry as a rainy, cool summer would be to beach resorts. As its neighboring ski areas, the Berkshire Mountains in western Connecticut were lacking snow. That's where one Walter R. Schoenknecht was trying to run a ski business. Schoenknecht would someday be the king of a Vermont mountain that was the site of the world's largest ski resort. When he first discovered it on October 2, 1946, it was known as Mount Pisgah. He opted for a more marketable name: Mount Snow.[11]

But in that snowless winter in the Berkshires, Schoenknecht was operating a modest rope-tow ski area. Necessity motivated him to opt for a page from Toronto, Bear Mountain, the Boston Garden, and Madison Square Garden.

Schoenknecht by then had leased state forest land to set up his Mohawk Mountain ski area for the winter of 1947–1948. The original complex included six rope tows and a base lodge.

The season was an overwhelming success, wrought by a harvest of snow. According to Nils Ericksen, who was technical editor of *Ski Area Management* magazine, on December 26, coincidentally 26 inches of snow fell on the resort, and all that was heaped atop a 12-inch base. The seasonal total was close to seven feet, and business was so robust that the following summer, he decided to add a building and four more rope tows.

By experience, he already knew that nature wasn't always going to be so bountiful with snow.

Before Mohawk, he and his wife had run a humble ski area on a leased property near Pittsfield, Massachusetts, equipped with all of one rope tow and three trails, one of which dropped toward a barbed-wire fence that the owner refused to remove. Not only did it lack for amenities, but also that first winter it so lacked for snow that Schoenknecht decided to surrender the lease.

So what happened at Mohawk shouldn't have been a shock to him. His good fortune ran into a serious dry patch in the winter of 1949–1950. The snow drought that season was so bad that it forced him to cease operations— but only temporarily. Schoenknecht wasn't held in the highest esteem by his rivals and critics, but no one could dispute his determination. In mid-January, Schoenknecht counteracted nature's parsimony by importing 700 truckloads of block ice from a site about 10 miles from Mohawk. He used a massive ice chipper to create a snow-like surface, although at one point, work was halted because someone had dropped tongs into the machine. By spreading the crushed ice, the trail was able to open. The frozen surface lasted only two weeks; however, he was able to sell 1,800 tickets in a single weekend.[12]

"I Know How to Make Snow"

That same winter would turn out be a watershed season for the skiing industry, and that had a lot to do with Mohawk and Schoenknecht, who was looking beyond crushed ice.

Three of the Mohawk customers were his friends: Art Hunt, Wayne Pierce, and Dave Ritchey. The three had formed a corporation, the Tey Manufacturing Co.—named for the last letters of their last names. They had set up shop in Milford, Connecticut, about 60 miles south of Mohawk, and not far from the coast, and they were making and selling aluminum skis. Not surprisingly, they discovered that selling skis wasn't easy when it wasn't snowing, that their business was frustratingly weather-sensitive. What if they could insulate their enterprise from the caprice of winter?

It happened that a revolutionary development in the history of winter sports was underway in both Milford and Mohawk during that same snowless winter. With a secrecy that would have been more befitting for the Manhattan Project, in December Tey delivered a snowmaking apparatus to Schoenknecht's Mohawk resort. The device, consisting of hoses, compressed air, and specially designed nozzles, emitted what was described as a "supersonic shriek." It also emitted showers of snow.[13]

The Tey partners had been experimenting with the process at the Milford plant, and testing continued at both Milford and Mohawk through the winter. By the end of that season, the partners were confident that they truly had the answer to snow deprivation. During the Northeast snow drought, the Tey Manufacturing plant became a white oasis. By mid-March, the factory was producing as much as 18 inches of snow a night. Wayne Pierce was confident that ski operators no longer had to wait for nature to take its course. He famously declared, "I know how to make snow."[14]

"The invention relates to a new and useful method of making snow." Thus began the U.S. patent application that Pierce would submit nine months later, on December 14, 1950. Pierce noted that the two preceding winters were "extremely poor and occasioned considerable hardship" on the ski industry. If a way could be found to manufacture snow, the "financial hazards" would be "materially reduced." He noted that ski areas had tried "dry" remedies to create low-friction surfaces, for example, pine needles and straw. He also referred to what his friend Schoenknecht had attempted by trucking in ice. That had proved effective, but aside from the time and cost involved, what Schoenknecht was using was not "snow." The application cited the government's technical definition of that precious gift—that is, precipitation in the form of small ice crystals. In other words, an ice cube was not a snowflake. In experiments, Pierce boasted, his compressed-air machine was able to transform two tons of water into snow at an air temperature of 25 degrees Fahrenheit. His proposal included several complicated illustrations followed by several pages of description not recommended for those who have a limited appetite for tedium. Yet, essentially, the original "Tey Gun" apparatus resembled a firehose mounted on a tripod. Pierce used a spray-paint compressor—the same one the company employed to paint those skis that didn't sell during snowless winters—a nozzle, and a hose.[15]

Orange Ice

Emboldened by the successes, the company cranked up its publicity machine, and articles later appeared in *Life* magazine and the *Wall Street Journal*. Meanwhile, the art of snowmaking was poised to prosper from an unlikely source—the Florida citrus business.

For aluminum-piping and other supplies for its snowmaking equipment, Tey had been approaching various irrigation companies. The one that offered the lowest price was an outfit called Larchmont Farms, operated by Phil and Joe Tropeano, in Lexington, Massachusetts. After Tey filed the patent application, the Tropeano brothers contacted Tey and suggested that with modifications one of its irrigation nozzles might be repurposed for snowmaking duties.

Larchmont had no direct involvement in the ski business; all it cared about was protecting oranges from killer frosts.[16] The irrigation nozzle it was offering emitted steam and water to arm the orange groves against those occasional cold air masses that spill all the way into the Sunshine State. It might seem counterintuitive, but spraying irrigated water on oranges before the temperature drops below freezing is a cheap and effective way to save the oranges. When the water turns to ice, the change of state releases latent heat. Then the ice that develops on the skins becomes an insulator.[17]

Phil Tropeano was convinced that the same basic nozzle technology could be used to help create snow. He put his hypothesis to the ground-truth test.

Three days before Christmas 1950, on a frigid Lexington morning, he trained his nozzles on the brown frozen dirt and fired away. The ground was brown no more, and Tropeano declared that the fresh, rich covering of whiteness indeed was snow. Once Larchmont accommodated Tey's request for the snowmaking nozzle, an industry-changing technology was on the runway. Takeoff, however, wasn't imminent.[18]

Water Guns

As stingy as nature could be with snow in the north country, it could be stingier to the south. Owners of ski areas in the Pennsylvania's Pocono Mountains had more experience with snow deprivation than they would care to recount. In a decent winter, the Poconos might be able to count on 60 or 65 inches of snow at higher elevations. That's about a half or a third of what the Green Mountains of Vermont receive on average. Thus, word of a promising remedy for barren slopes must have been received with tremendous enthusiasm at Pennsylvania's first ski resort, Big Boulder, built at the site of a former Lehigh Coal and Navigation Co. retreat near Lake Harmony, not far from anthracite coal country.

After suffering through a painful version of the snow droughts that also plagued its neighbors to the north, Big Boulder took a chance and purchased a system from Tey. Its debut during the 1950 Christmas season wasn't a roaring success, although it was at least roaring. The guns that were supposed to disperse snow instead spit out water. It must have been a particularly stinging disappointment for John Guresh, at the time a Big Boulder manager who already had a long association with the ski area, even before its beginnings.

"It seemed hopeless," he recalled on the 25th anniversary of the fiasco. "We just didn't have the know-how." On that day in 1950, a trickle of water is not something the former Lehigh Coal water boy wanted to see, but Guresh would become an unsung hero in the history of snowmaking.[19]

Guresh went to work for Lehigh Coal as a teenager and graduated from water boy to skilled mechanic to bulldozer operator. He cleared the road that would lead to the resort and helped build the lodge at the base of Big Boulder. He became the construction foreman and worked as mountain manager during the winter.[20] The failure of the supposed snow guns had him flummoxed.

By the end of the winter Big Boulder's operators decided to trash the system for which they had paid $80,000, about $800,000 in 2020 dollars.[21] "We sold it for scrap," Guresh said. "We lost everything."[22]

Meanwhile Schoenknecht was crowing about his snowmaking success at Mohawk. As the calendar turned to 1951, he announced publicly that he had added three inches of snow to an eight-inch base at Mohawk and that he planned to make snow every weekend night.

Guresh was sufficiently intrigued to visit Mohawk to have a look at its snowmaking system, but the trip turned out to be a waste. He picked a bad time; the weather was unseasonably warm. "The wait for colder weather was futile and boring," he told a *New York Times* reporter. "Snow? All most of us got was drunk."[23]

Despite Schoenknecht's boasts, snowmaking technology confronted mountains of skepticism regarding whether it was worth the trouble. Said one Vermont ski operator, "The best snow is the type that falls from the skies because it is the cheapest kind."[24]

But neither the Pocono setback nor any others were going to stop the inevitable. The 1950s would constitute a heyday for the manufacturing economy of the United States, and one product that snowballed in popularity was "snow."

In the winter of 1952–1953, the ski area at Grossinger's Hotel, in Liberty, New York, became the site of what was considered the first major snowmaking project in the country.[25] It was after that winter, in May, that Schoenknecht took a ski-jump leap toward realizing his outsized ambitions, forever changing the public's vision of the "ski area." He persuaded a farmer in the town of Dover, Vermont, to sell him 500 acres at Mount Pisgah for $15,000, approximately $150,000 in 2020 money. All that farm equipment would one day be replaced by the "most powerful snow guns in the Northeast." Fortuitously, the farmer's name was Reuben Snow, a natural name for Schoenknecht's mountain.[26]

It was undeniably true that snow from the skies was far and away the best option for the ski mountains, but it was equally undeniable that in some winters the skies simply wouldn't give it up. Some years they would have to make their own—especially in the East in areas south of the Green, White, and Adirondack mountains—hiccups notwithstanding. And those hiccups were inevitable.

The winter of 1956–1957 was another of those watershed years for the machine era. In November, Boyne Mountain announced that it planned to cover a 2,000-foot-long area with a 490-foot vertical drop with artificial snow, with an Austrian ski star on hand to conduct a clinic with five instructors.[27]

In Massachusetts, the Bousquet Ski Area, in Pittsfield, crowned a 1,500-foot slope with machine snow but not without adventure.

John Hitchcock, a ski instructor at Bousquet and a writer for the *Berkshire Eagle* at the time, told a wonderful story about the early days of snowmaking at Bousquet. Donald Soviero, who ran Bousquet, had hired Larchmont to set up a system powered by diesel air compressors. Hitchcock was present on November 30, 1956, when Soviero prepared to demonstrate the device to an audience that included several New England ski operators. When he turned on the compressor, however, an explosion sent the network of aboveground aluminum pipes skyward.

The Bousquet crew managed to reassemble the pipes and get the system working; however, witnesses were not impressed by what appeared to be a labor-intensive, expensive process that didn't produce much in the way of snow.[28] As it turned out, the 1956–1957 season was one of those nonwinters.[29] A strong El Nino event, that anomalous warming of tropical Pacific water that generates warming winds from the west that can flood North America with warmth, was ripening. During El Nino, winters in the East can go begging for snow, especially in the first half of the season.

By Christmas 1956, even the normally snowiest slopes of Vermont were bare. Not at Bousquet. Having recovered from that explosion, Bousquet kept the snow machine operating, and Hitchcock recalled that Russell Slope was so crowded it was hard to discern movement among the throngs. *Life* magazine sent a photographer to capture the scene, and the snowmaking virus was unstoppable.[30]

That same winter 18 nozzles were spewing snow atop a quarter-mile slope, rising 246 feet, on Laurel Mountain, in western Pennsylvania, and on the other side of the state, John Guresh was newly confident that he could cover Big Boulder's slopes with machine-made snow.

Try, Try Again

After the crushing disappointment of the first attempt at snowmaking at Big Boulder, Guresh left the resort and took a maintenance job with the Pennsylvania Turnpike. But Big Boulder wasn't done with John Guresh. In 1955, it decided it wanted to take another shot at a cure for barren slopes and persuaded Guresh to come back to try to get the system working. This time it tried a Larchmont system.[31]

According to a history published by the Pennsylvania Snow Sports Museum, Guresh "would often tinker with the machine for an entire night, go home for an hour's nap, and return to the mountain." While his colleagues might have found his diligence admirable, they wondered about his sanity.

Through trial and error, he eventually discovered the right recipe of water and air, then located sufficiently powerful compressors. And at long last he believed he had solved a key problem—the freezing of the water lines. "One of the biggest problems," he said, "was keeping the water moving fast enough to prevent freezing before it could be sprayed in microscopic beads in the atmosphere."

He developed a circulation system that prevented the freeze-ups.[32] After a year of tinkering, the system was ready for launching.

In the winter of 1956–1957, a device that looked like it was impersonating a lawn sprinkler dispersed crystals at Big Boulder and did a convincing job of impersonating real snow. Guresh was generous with his knowledge and shared his findings with the Tey company, maker of that 1950 dud machine. Tey was so impressed with his expertise that it tried to recruit him to work for the company; however, he decided to stay at the place that he had helped to build. Throughout his career, Guresh appeared to show no interest in wealth. He helped other ski resorts with snowmaking systems. He helped with the installation of snowmakers at Camelback and Elk Mountain, two popular Poconos resorts. His concepts were employed by other ski areas throughout the nation, but he never attempted to patent or capitalize on his refinements. He later was honored by the Pocono Mountain Vacation Bureau as the "originator of the snowmaking business," and by the governor of Pennsylvania.[33]

For all that, he not only failed to reap riches from this work on the machines, but also personally didn't get to experience the results. He was unable to ski because of a bum knee.

By the end of the decade, 18 ski areas in New York and New England were listed as having snowmaking systems in the Eastern Ski Area Directory. Not every resort was buying into the program; Killington, for example, did not install a system until the 1960s. Not surprisingly, the business diversified as it underwent a natural growth spurt. Tey decided to sell Pierce's patent to a company that eventually was bought by Larchmont, which would have a long career in the industry.

Being essentially a compressed-air and water machine, the primitive Tey Gun was uproariously noisy and a model of noisy energy inefficiency. That inevitably would change. In 1961, at onset of a decade of cold and snowy winters in the Northeast, a patent was issued for a fan snowmaker that required no air compressor.[34] The inventor, A. W. Hansen, resorted to a 90-horsepower Continental airline engine, which activated a propeller. Taking a page from the atmosphere, he noted in the application that "dirty water" and crushed ice helped to produce condensation nuclei that promote crystal formation.[35]

As the 1960s progressed, ski areas would have dozens of options for airless and air/water machines, but some in the industry remained reluctant to embrace the new technologies. Among them were two popular names in the Vermont ski business, Bromley Mountain and Magic Mountain. When Bromley's Fred Pabst gave in, he went all the way, installing a snowmaking system with a $750 million price tag, or about $5 billion in 2020 dollars. He declared it the "world's largest" such system. "No one thought you could cover a whole damn mountain with snowmaking," he said.

"I held out because it's so damned expensive to buy that snow equipment and keep it running," Pabst told a reporter years later. "But I've found it's more expensive to have to close our ticket windows because we're out of snow." Magic Mountain's Hans Thorner, the one who famously said, "The best snow is the type that falls from the skies," also caved. "I had no choice," he said. "I could accept snowless winters, but I couldn't face up to having skiers rushing down nearby mountains when only wind was rushing down mine."[36]

In 1974, Killington, one of the nation's most popular ski resorts, decided to crank up its snow factory, committing to spreading machine snow at the higher elevations and covering 154 acres of trails. It takes 4.1 gallons of water to manufacture a cubic foot of snow, Blue Mountain's Bob Taylor told me. In the winter of 2018–1919, another one that wasn't particularly snowy, the ski area used 240 million gallons of water for snowmaking.[37] Keep in mind that to cover an acre with an inch of snow requires about 12,000 gallons of water. Part of Killington founder Pres Smith's reasoning was that skiers were becoming more proficient, and that meant more wear and tear on the slopes. Smith predicted that skiers would demand machine-made snow.

It took the poor snow season of 1980, when only two Vermont areas finished in the black, to convince many operators of the necessity of snowmaking as a means to providing a "consistent product."[38]

The winter of 1979–1980 was a strange one. It was an upside-down season, with Norfolk, Virginia, receiving more snow than Burlington, Vermont. All but two Vermont ski areas finished in the red.[39] It might have been an economic disaster for the resorts, but it was a boon to the future of snowmaking. At the 1981 annual meeting, a Killington official declared, "The traditional assumption that it's going to snow and then expressing disappointment and negative impacts when it doesn't is a thing of the past."[40]

By the mid-1990s, more than 1,200 snowmaking systems were operating in the United States and 19 other countries. These days it might be impossible to imagine that once upon a time not so long ago, ski resorts relied solely on nature. These days it might be impossible to imagine a ski resort not relying on manufactured snow. Snowmaking has become an absolute necessity in the East. One Pennsylvania resort operator was quoted as saying, "A ski area is nothing but a great big machine."[41]

This ever-expanding ability to create snow has forever transformed the mountains of the East and West—physically and culturally. Ski trails would require removing trees and other dense vegetation, contributing to erosion. A ski resort requires massive infrastructure development and construction in areas previously undisturbed, except perhaps by hikers. Ski trails aren't surgical cuts. They require extensive removal of trees and other vegetation. That ability to transform the landscape evolved into an ability to transform the very definition of ski resort. Today ski mountains aren't just for winter anymore; they are year-round playgrounds. At such places as Wildcat and Loon in the White Mountains, you can ride ziplines in July. Walt Schoenknecht had a lot to do with all that.

From Pisgah to "Snow"

Schoenknecht not only was a pioneer in the career of snowmaking, but also his role in shaping the public's expectation of the skiing experience remains legendary. From that humble rope-tow ski run at Mohawk Mountain, he went on to execute dreams that some viewed as nightmares. "Perhaps his vision . . . would turn an environmentalist pale," I. William Berry wrote in his letter nominating Schoenknecht to the U.S. National Ski Hall of Fame, "but it would make a skier sing."

By Berry's reckoning, Schoenknecht had delivered the ski business from a dark age in which skiing wasn't the pursuit of a good time. "Before Walter, skiing was hardship and survival," he wrote, with "rough-hewn base-lodges heated by potbelly stoves."

Schoenknecht, he said, made "skiing fun for all levels of skiers. Take two lessons and you could sweep down the mountain from the top." Berry's prose was a match for Schoenknecht's extravagances. Berry wrote of "big, colorful, two-people gondolas . . . like hatching Easter Eggs from clock-tower terminals."

With the ability to make snow, the trails no longer had to face due north to hide from the sun. You could ski "warm and loose and happy, basking in the sun on those east-facing runs. . . . Everyone laughed at the excesses, the goldfish and the Oriental pools and the archaeology of lifts, and they came back again in an avalanche because it was fun."

Yes, the remaking of the mountain had drawbacks. "The purity of an earlier time took a beating," Berry said, "but the sport of skiing grew mightily in the '50s and '60s nowhere more joyously than at Mount Snow."[42]

Remarked one of Schoenknecht's critics, "Come to Mount Snow, . . . see his Disneyland and Coney Island of the Snow Belt."[43]

Mount Snow had opened in time for the abundantly snowy 1954–1955 season with two chain-driven lifts, five trails, and two slopes. Schoenknecht added chairlifts in subsequent seasons, and in the 1958–1959 season, he made quite a splash. The resort installed a kidney-shaped swimming pool with waters heated to a Jacuzzi-like 96 degrees.[44] It wasn't cheap or energy efficient. The heating plant was consuming 6,000 gallons of fuel a week. The expense was worth it if it drew the customer. "A skier won't let you stand still," Schoenknecht told a Sports Illustrated interviewer. "You must provide something new, something tremendously exciting every year to get him back."

That, he did.

In that 1961 article, Schoenknecht foresaw a fantastic future ending in a galactic collapse, followed by rebirth. In 25 years, he predicted, "we'll have poured about $75 million into this place, and that will be the time to start tearing down. We'll tear it down and we'll start over. Everything will be brand-new again. It'll be weird and tremendous, and second to none. It will be absolutely fabulous."[45]

It didn't quite happen that way. The capital expenditures and energy costs led to inevitable financial problems. Schoenknecht merged his enterprise with another company in 1971, and a few years later he left Mount Snow for Mohawk.

After the energy crisis of 1973–1974, and a sequence of down winters, Mount Snow went into bankruptcy in 1975, and eventually was purchased by the Sherburne Corp. in 1977.[46]

But variations of the Schoenknecht vision have survived in mountain areas throughout the country, and they have come with environmental costs.

Slippery Slope for the Environment?

The proliferation of ski areas has led to thousands of miles of road beds plowed through once-pristine mountainous and forested preserves to make way for carbon-emitting, polluting vehicles, some of them towing carbon-spewing, polluting snowmobiles. The snowmaking material itself has been a source of contention, particularly in the water-challenged West. Several resorts have made agreements with local government to use treated wastewater, and that is at the heart of a bitter controversy in Arizona.

The spectacular Arizona Snowbowl resort, seven miles outside Flagstaff in the beautiful San Francisco Peaks, on average receives about 260 inches of snow per year, according to its site. Yet, given the variability of its winters and ever-increasing anxieties that worldwide warming is trimming the season at both ends, it relies on supplemental snowmaking. It uses machines to nourish two-thirds of its 55 trails, which accommodate skiers, snowboarders, and, more frequently these days, ski-bikers. And they do this in a region where the mountains do an efficient job of intercepting moisture, depriving lower elevations of rain and snow, a typical precipitation profile in the West. Snowbowl features Alpine skiing and is the site of Mount Agassiz, named for the man who is credited with explaining the mystery of glaciation. In 2012, it became the first resort in the country to use wastewater for snowmaking.[47]

What *U.S. News* has called "one of the state's best spots for skiing and snowboarding" was once something quite different for the Hopi Indians. It was a sacred place. The Hopi frequently traveled to and through the area. At the base of the slopes is what the Hopi considered a "spiritual birthplace." Thus, the resort's decision to use treated wastewater prompted bitter protest from the Hopi, along with other tribes and environmental groups. They filed suit to stop Snowbowl on several grounds and argued that the use of wastewater constituted a "public nuisance." The litigation worked its way to the Arizona Supreme Court.

"Prevailing winds will blow the snow well beyond the boundaries of Snowbowl, covering sacred land, shrines, springs, and other natural resources with the reclaimed wastewater," one of the justice's wrote.

> In the spring melt, sacred springs will be tainted with the melting wastewater, turning formerly pure ceremonial locations into a secondary sewer. Moreover, the myriad chemicals in the water will wreak unknown damage on the local

ecosystem, further degrading traditional and sacred Hopi resources and locations. In sum, the Hopi face the destruction and desecration of some of their most sacred locations and practices.

That, however, was a dissenting opinion. In its November 2018 decision, the court dismissed the complaint and upheld Snowbowl's right to use wastewater. The majority noted that Snowbowl had been purchasing the wastewater from the city of Flagstaff for 16 years and that the environmental issues had been vetted extensively. Since the San Francisco Peaks are on federal lands, the U.S. Forest Service had conducted a "lengthy environmental impact inquiry, culminating in that agency's approval" of the practice. Furthermore, it held, "Environmental damage to public land with religious, cultural, or emotional significance to the plaintiff is not special injury for public nuisance purposes."[48]

While environmental concerns surrounding the winter-sports industry have persisted for decades, it continues to prosper. Environmental tensions are inevitable. In 1970, worries concerning how ski areas might be polluting water persuaded Vermont to pass tough environmental laws that actually halted ski area expansion.[49] But not for long.

Since anyone began keeping track of human activity, snow has always inspired an element of play, with a magical power to transform adults into children, albeit expensively outfitted ones. The growth of the winter-sports business was a speeding toboggan that was not to be stopped. "Years ago you went out and hooked up hoses and turned on valves," said Taylor, who has worked at Blue Mountain for 30 years. He opened a control box that looked like a computer's central nervous system. "Now we have to be able to fix all this (expletive)."[50] But the process itself is rather basic. Compressed air or high-powered fans eject the moisture into the air, where it condenses into hexagonal snow crystals, assuming the air is willing. Snow can form at temperatures above freezing if it is dry enough. The drier the air, the more evaporation, which has a cooling effect not unlike that of sweat when it vanishes from the skin. The "wet bulb" temperature indicates how cool the air will be if it evaporates all the moisture, and Taylor said that as long as that wet bulb reading is 28 or lower, he can make snow.[51] He could do it with a temperature of 37 if the relative humidity was 30 percent or less.

Resorts do count on real snow to fall in the places where their customers live, especially in the urban centers on the coastal plain, where they try to get rid of the stuff rather than make it; snow in the cities is an ultimate marketing tool for the ski areas, like the image of a cold Coke during a heat wave.

Still, skiers and even honest resort operators will tell you that on a mountain slope, there's nothing like the real thing.

Snowmaking versus Snow-Faking

For whatever flaws in the ski industry's snowmaking technology, it has been miles and miles ahead of moviemakers, who have struggled with "snow" since the cameras started rolling. Disney's decision to draft Caltech's Ken Libbrecht to consult on snow matters during the production of *Frozen* was a refreshing counterpoint to the Hollywood tradition.

The environmental issues associated with the use of wastewater for snow-making are almost trivial compared with the questions raised by the use of another material to make fake snow in movies: asbestos.

Along with its splendidly idiosyncratic characters, magical music, and choreography, *The Wizard of Oz* was a triumph of special effects. One of the most memorable is the poppy-field scene, which required the planting of 40,000 artificial flowers on the MGM set. Dorothy was under a spell that was dispelled by Glinda the Good Witch of the West, who crowned the scene with snowflakes. Glinda no doubt was unaware that she was dispensing toxic flakes of chrysotile, more commonly known as asbestos. It would one day be identified as a clear and present health danger, strongly linked to mesothelioma, a debilitating lung disease.

In *The Gold Rush*, actually shot in the Sierras near the California–Nevada border and released in 1925, Charlie Chaplin resorted to combinations of salt and flour to affect the snowscape of gold-rush Alaska. But until about 1930, movies generally used variants of cotton for wintry scenes. Cotton presented an obvious hazard: It was quite flammable. So movies turned to a fireproofing material, asbestos. As the Mesothelioma Lawyer Center reminds visitors to its site, it was sold popularly under the brand names "Pure White," "Snow Drift," and, of course, "White Christmas."[52]

The explanation for how snow and Christmas became inextricably linked in the United States is beyond the realm of the logical and a process that began unfolding in colonial days. But two 20th-century figures who had prominent roles in cementing the concept were a Jewish songwriter and a golden-voiced crooner. Bing Crosby first performed Irving Berlin's "White Christmas" on a radio broadcast and reprised it more famously in the movie *Holiday Inn*, a romantic comedy set in an inn in Connecticut. It first appeared in theaters in 1942, and has been recycled by TV movie networks and stations each year since. In the climactic scene in which he croons the tune to his mesmerized love interest, naturally snow is falling—or rather, unnaturally. In the annals of fake snow, the stuff is about as counterfeit as it gets. Even in my grade-school days I was put off by the fact that so many snow scenes in movies occurred with no evidence that it was in any way cold. Never would I see vapor emerging from the mouths of actors or actresses when they were talking, for example. In *Holiday Inn*, it is clear that the snow is falling on a movie set. What moviegoers (nor the cast, in all probability) didn't realize was that yet again, the "snow" was asbestos.

Asbestos was considered for use as a snow surrogate in the 1946 Frank Capra classic *It's a Wonderful Life*. Fortunately for Jimmy Stewart, Donna Reed, and colleagues, Capra and special-effects wizard Russell Sherman invented an alternative. The dangers of asbestos had not yet been fully revealed, but one of the major nuisances for popular snow surrogates, including white-painted Cornflakes, was noise. The crunching sounds made by the footsteps of the actors on the fake snow were so loud that they could obscure the spoken words.

Sherman and his crew concocted a blend that included the flame retardant Foamite, used in fire extinguishers, sugar, and water. The product was a sprayable material, and several thousand gallons of it were applied to give the fictional town of Bedford Falls a romanticized covering of snow. The Foamite muted the footsteps to such an extent that Capra was able to record the dialogue as it happened. For the snowy efforts, the Motion Picture Academy bestowed a special effects award on the movie. As much as I have enjoyed the movie throughout the years, the fact that the snow job was honored by Hollywood is absolutely incredible to me.

For convincing snow cover, I would put *It's a Wonderful Life* in a league with *Holiday Inn*, and the falling snow was even more unrealistic. Once again, it's hard to believe it's snowing in the bleak midwinter when not a trace of vapor is emitted from the actors' mouths. In fact, in the scene where the George Bailey character, played by Jimmy Stewart, jumps into the presumably freezing waters on Christmas Eve to rescue "Clarence," his guardian angel, Stewart is sweating.

As an article in *Life* magazine pointed out, the movie wasn't shot in the bleak midwinter, but in the summer of 1946, and not in "Bedford Falls" or any place on Earth where it might be snowing on December 24 or any other day. It was filmed during a sequence of days with 90-plus-degree temperatures on RKO property in Encino, California. Capra told the cast and crew to take a day off during the filming to get a break from the heat.[53]

Perhaps it is fitting that movie snow in those Christmas movies should strain credulity for anyone who pays attention to weather and cares in the least about actual snow. On any given Christmas, bare ground is the rule for an overwhelming majority of Americans. Yet, American movies that have Christmas scenes will have at least a dash of snow, fake in all likelihood. The only exception that comes to mind is A Christmas Story, which has actual winter scenes shot in a cold location. But even the folks who made that one couldn't leave it alone. On Christmas morning, Ralphie wakes up to find that a requisite wonderland of snow has landed upon his backyard—and looks about as real as the snow in It's a Wonderful Life.

What is behind America's fixation with snow at Christmas?

One person who dedicated time, energy, research, and intelligence to that overwhelming question was the late Tristram Coffin, a University of Pennsylvania folklorist and author of The Book of Christmas Folklore. I had the good fortune to speak with him a number of times before he died in 2012, and he was generous with his time and insights.

Coffin believed the seed of the concept was planted in the 4th century, during the reign of Emperor Diocletian, and ripened in the United States in the 19th and 20th centuries. The tradition of "Santa Claus" was rooted in the legacy of the bishop of Myra, Nicholas, who died in 342, and later became St. Nicholas. It is believed that he wore red robes—although keep in mind that Coffin was piecing together information from 17 centuries ago—as depicted in images. The garments, by the way, appear more ecclesiastical and tasteful than the ones worn by the average department-store Santa, but, yes, they are a reddish color. St. Nicholas became revered in Northern Europe, particularly in the Netherlands. On the anniversary of his death, December 6, Dutch legend had it that he would appear in the sky on a white horse, and children would leave hay in their wooden shoes for the steed.

The Dutch carried the tradition with them when they settled in New York Dutch country, and writer Washington Irving popularized the image of "Santa Claus" in a satiric poem. But it was another poem, credited to a professor of Greek and Oriental literature named Clement Clarke Moore, that bound snow with Christmas with an immutable epoxy.

The formal title of the poem was "An Account of the Visit of St. Nicholas," and its authorship remains a subject of debate. It first appeared in 1836, in the Troy, New York, Sentinel. Most people know it as "'Twas the Night before Christmas," for one of the most famous first lines in all of literature. To the St. Nicholas narrative, the poem added reindeer and a sleigh, the equivalent of a 19th-century pickup truck brimming with gifts. Sleigh blades need a friction-free frozen surface as much as 21st-century tires need macadam.[54]

Troy, in the Upper Hudson Valley, not far from Albany, is one of the places where a snow cover at Christmas is a reasonable possibility. Those "clipper" systems barreling eastward from southwestern Canada sometimes pass over that area, and the region is close enough to the Atlantic for nor'easter impacts, yet far enough away to evade the warming impacts of onshore winds that change snow to rain along the coastal plain. And, in 1836, a Christmas snow cover would have been all the more likely. That was the twilight of the "Little Ice Age," an era dating to the Renaissance when Europe, northeastern North America, and perhaps other parts of the planet were significantly colder than they are today.[55] Another rather famous individual writing in that era was one Charles Dickens. In A Christmas Carol, snow adds more than a dash of atmosphere.

The Christmas–snow connection was forever solidified in the 1940s with Crosby's rendering of "White Christmas" in Holiday Inn. It so happens that in the climate records, the 1940s mark the beginning of a worldwide warming spell. As we've seen, however, no amount of warming could stop movie snow.

The warming continued into the 1950s, and as a matter of record, it had no effect on the snow in the movie that sealed the deal between Christmas and snow in America: White Christmas.

CHAPTER TEN

The Vanishing

The Future of Snow

The movie *White Christmas* is a classic American fairy tale built on that romanticized ideal of snow and Christmas. Snow cover is common in some parts of the nation on December 25—the higher mountains, the northern tier, the New York Snow Belt, the Upper Midwest—but is the exception in the densest population centers. Florida, the California coast, and most of Texas are unlikely venues for dashing through the snow. The Northeast Corridor certainly has had more than its rations of early winter storms, but bare ground is more likely than snow cover around Christmas, even in the Boston area, where the chance of waking up to an inch or more of snow on the ground on December 25 is less than 20 percent, according to the data assembled by the National Centers for Environmental Information.[1] Atlantic Ocean sea-surface temperatures typically are in the upper 30s to low 40s, and since nor'easters can stir up onshore winds throughout the atmosphere, what starts as snow doesn't always stay as snow.

Not that *White Christmas* was intended to serve as a treatise on meteorological realities. From the opening moments when a 50-year-old Bing Crosby, starring as "Captain Wallace," sings the title song and then is saved from a falling wall by PFC Davis (Danny Kaye), the movie is as bereft of reality as it is a stuffed stocking of music and dancing. In the States, Bing and Danny team up and, through a variety of implausible circumstances, end up at the "Columbia Inn" in "Pine Tree," which happens to be owned by a dejected and forgotten General Tom Waverly, who appears to have serious delusional issues, based on the analysis of a psychologist friend of ours with whom we have watched this movie at least 25 times; please don't ask why. Those who had seen *Holiday Inn* might recognize the Columbia, given that it's the same building, only this time it's in color, and located in Vermont rather than Connecticut.

One of the more incongruous moments has Bing crooning "Count Your Blessings" to an enraptured Rosemary Clooney, a song in which he mentions admiring his sleeping children in the family nursery. That's rather disorienting given that until this point, a linchpin of the plot has been Bing's utter immunity to family life, although in the presence of Rosemary Clooney, it is melting faster than a snowpack in a heat wave. Details, details. Then we have Bing singing a song lamenting the problem of unemployment among postwar generals. The movie was released in 1954; the president of the United States was former general Dwight D. Eisenhower. Through it all, Bing is that too-good-to-be-true selfless American hero who manages to marshal his entire musical revue to perform at the inn for the psychological and financial well-being of General Waverly. In the over-the-top ending, we see the stars dressed in dazzling red Santa sets singing—you guessed it—as a set of barn doors open to reveal some of the phoniest-looking snow in the history of fake flakes.

The movie does have one realistic moment, and it comes at the end of that most-improbable plot twist in which Wallace and Davis end up riding a train from Florida to Vermont right before Christmas. The train arrives to a scene of bare ground and snowless tree branches. At the station they encounter frustrated skiers who have decided to pack up and head for home or elsewhere. The boys are informed that it hadn't snowed since Thanksgiving.

In most years, the Green Mountain ski areas can rely on a more-than-decent snow cover in mid- and late December, but snow-starved Decembers even up that way aren't all that unusual, a fact that was the impetus for the snowmaking pioneers. Recall that the winters of 1948–1949 and 1949–1950, when U.S. snowmaking technology was born, were distressingly snow deprived for the operators of ski areas in the Northeast.

What if those winters were to become the norm as a consequence of rising worldwide temperatures? *E-magazine* addressed the question with the headline, "Is Skiing Dead Due to Global Warming?"[2]

Beyond skiing, in the West—so dependent on winter's bounty in the mountains for its water—snowless winters would be a certifiable disaster. But a permanent snow drought could be a devastating loss to the national economy. And what if it became too warm even to make snow with any regularity?

In a 2009–2010 survey, a pair of University of New Hampshire researchers estimated that the winter-sports industry generated more than $12 billion in economic activity, which included money spent on hotel rooms, gas, groceries, and, of course, drinking. They calculated that more than 20 million people participated in winter-sport activities and that federal, state, and local treasuries harvested $3.1 billion in tax dollars. All this skiing, snowboarding, and snowmobiling helped support almost 212,000 jobs.[3]

Calling Dr. Snow

Thus, concerns about the state of winter in the United States are more than recreational, and these days the unthinkable is being articulated: Are we seeing the end of snow? It would be death to the weather forums that provide hope and group therapy for fans of snow throughout the country and feed the initiable desires of what our son calls snOwCD.

As a lover of snow, Dave Robinson lives and works in the heart of the heartbreak-euphoria zone. Northern New Jersey, a populous annex of the New York city-state, has had its ration of mega-snowstorms, particularly in the 21st century, and way more than its share of just-misses as nor'easters over the Atlantic bomb out far enough north to shut down Boston and frustrate the snow-philes from New York City on south or develop too far south to export snow past Virginia. When the "big one" snubs a snow-lover, those harsh north winds that chase the cloud debris and are the parting gift of storms heading into the far North Atlantic have a particular sting.

At the peak of a February storm during the amazing winter of 2009–2010, I called Robinson at his house, and I wish I had left him alone. He was distracted, I could tell, the way you know that someone with whom you're speaking on the phone has a mind disconnected from the voice. At the moment, he told me, he was experiencing the heaviest snow he had witnessed in his lifetime. I should have left him alone to luxuriate.

But I was bothering him for a reason. I was on deadline, and if you're going to write about snow, he is at the very top of the expertise list.

Dave Robinson, a "distinguished" professor of geography at Rutgers University, is the nation's longest-serving state climatologist. He is also Dr. Snow, and under his stewardship the unlikely venue of Piscataway, New Jersey, has become an international snow nexus. Robinson is an unassuming man with a well-measured sense of humor. When the school newspaper identified him as a professor of "geology," he accepted it unresentfully, an inside joke he would live with. He takes snow quite seriously. He has become the National Oceanic and Atmospheric Administration (NOAA) and the Intergovernmental Panel on Climate Change's go-to person on snow cover, which is becoming evermore important.

Robinson describes himself as the accidental polar scientist. He grew up in Tenafly, not far from where he now works, and, more significantly, just five miles from the Palisades on the banks of the Hudson River on the Jersey side. That's where Columbia University's storied Lamont-Doherty Observatory is located.

By the time Robinson enrolled in Dickinson College, in Carlisle, Pennsylvania, he had already long been captured by snow. His very first weather memory is Hurricane Donna, which blitzed Tenafly in September 1960, when he was five, and closed his kindergarten. Yet, what "captivated" him wasn't Donna, but the snows of the subsequent winter, when his town experienced three storms of 10-plus inches, something that wouldn't happen again until that winter of 2009–2010.

As an undergrad in 1976, he applied for a summer position at Lamont-Doherty, and from there his career was paved with snow. By luck, a Dickinson professor had an adviser at Lamont named Wally Broecker, a legend in the climate community whose research on Atlantic Ocean currents was the basis of that cli-fi disaster movie *The Day After Tomorrow*, the one in which the Gulf Stream stops and snow buries the Statue of Liberty. Robinson sent his application to Broecker's secretary and got hooked up with a professor who had just received a grant to study NOAA's satellite snow maps. The professor liked his work enough to invite him to apply to Columbia for grad school.

Robinson became quite the master at interpreting the satellite maps, which dated to 1967. After he became Dr. Robinson, he took his expertise with him to Rutgers, where he became chairman of the geography department. One thing he discovered was that NOAA didn't particularly care about the precision of the maps; the agency wasn't interested in building a climate record, but rather using the images for operational forecasting. Knowing the snow boundaries is important in predicting day-to-day weather; by refrigerating the overlying air, snow can influence the movement of upper-air winds that govern weather. Piecing together the disparate puzzle pieces, Robinson assembled a picture that he found quite astonishing and disturbing. Snow cover was disappearing ever sooner during the North American springs. He published a paper in 1990, referencing global warming.

But something about the maps was bothering him. "We realized there was an inconsistency with the way NOAA generated their numbers," he said. "It didn't change the results, but it altered the database. So we figured out what the problem was and then generated a database that was consistent." Robinson became *the* source for all matters of snow cover. The Intergovernmental Panel on Climate Change reports on snow rely on Robinson's analysis. He created the Rutgers Snow Lab, the climate community's go-to source for tracking snow and ice cover derived from satellite data. He posts reports daily. Even NOAA has relied on him. "That inconsistency by NOAA was the best thing that ever happened to me," he said.[4]

For a snow-lover, *The Case of the Missing Snow* would not be a mystery story that you'd care to read, let alone write.

Given his passions, and his expertise, what Robinson is seeing these days is at once intriguing and troubling. In a career of watching snow, he is not unlike an avid sports fan who has known the exhilaration of celebration, and the grief of watching his team lose the championship game. What he sees in the springs of recent decades suggests his team is losing. But almost nothing comes tidily in atmospheric research. Said Robinson, "The more you know, the less you know."[5] While snow boundaries have been retreating earlier, paradoxically, the snow cover in October and November in the Northern Hemisphere increased robustly from 2009 to 2019.

In the grand scheme, however, he is convinced that worldwide warming is eating the snow, and it's possible that the snowfall harvest is related to worldwide warming. Despite the autumnal snows of 2019, snow cover was below normal in eight of 12 months that year, and among the top 10 lowest in the period of record in four of them. Robinson maintains the world's longest continuous satellite record, but the period of usable observation dates only to 1967. It is likely that those 2019 figures would ring even more alarm bells if compared with the entire, colder 20th century.

A Warmer World

These days that the world has become warmer is indisputable. The evidence is in the global surface-temperature bases and further verified by the satellite data assembled by NASA scientists John Christy and Roy Spencer. In the environmental-activist community, Christy and Spencer have been branded as pariahs and "deniers." Yet, from what I have seen of their data, collected from satellite microwave measurements of the lower six miles of the atmosphere, the trends have tracked reasonably well with the surface datasets kept by the U.S. government's National Centers for Environmental Information (NCEI).

NCEI publishes monthly and annual comprehensive reports on the global temperature and the state of the climate. The annual reports typically garner some media attention; the monthlies typically don't. The media coverage usually is confined to the highlights—the third-warmest month/year on record—and what the reports say about the actual global temperature. In

2019, for example, NCEI reported that Earth's temperature was about 1.71 degrees Fahrenheit—give or take 0.27 degrees, and that's an important quali- fier—above the 20th-century average.[6] But what exactly does it mean to say Earth was 1.71 degrees above that average in a given year? And a reasonable, casual observer might wonder why on earth would such a puny-looking dif- ference make any difference whatsoever to the world's snow cover?

It's a shame those reports don't receive more detailed attention because they are full of fascinating information and assembled at great effort. The element that gets the most buzz—the actual overall change in the planet's surface temperature—is among the less interesting tidbits for my money. And I'm not sure what most of us think about what it means when the gov- ernment or another measuring entity says Earth's temperature was such and such degrees above "normal" or "average," or, more fundamentally, just what the global temperature *is*.

I learned from those who do this for a living that arriving at the number isn't quite as simple as sticking a thermometer into the mouth of the world and asking it to say, "Ahhhh." This requires some detective work.

NCEI's Deke Arndt—and as Robinson is the go-to person on snow cover, I would consider Arndt his counterpart on global temperature—explains that the published figures are the averages of averages, that is, the aver- age monthly temperatures at more than 2,000 stations from 180 countries throughout the world. They are daily averages—highs and lows in all the months' 24-hour periods divided by two—all divided by the number of days in a given month. Reports arrive promptly from some stations and not so promptly from others.[7] Some economically struggling nations have higher priorities than taking temperature measurements. When the Taliban shut down that Afghan Meteorological Office in Kabul, 100 years' worth of cli- mate records were destroyed.[8]

Ideally, ambient temperatures should be taken in carefully sheltered environments about six feet above ground level by carefully calibrated in- struments. Not all observing sites are created equal or maintained equally. Remember, first of all, that about 70 percent of Earth's surface is covered by oceans, and NCEI counts on reports from ships and buoys. Earth's land surface is splendidly diverse, with majestic mountain ranges alternating with valleys and expansive plains, jungled rain forests and ice sheets, grand fresh- water lakes holding mirrors to the skies, deserts, and, of course, the oceans, which are hardly placid. The varied surfaces have their own special relation- ships with incoming solar radiation, retaining and reflecting it differently.

NCEI uses various methods to compensate for weaknesses and discontinu- ities in the observations. That's why the monthly reports come with margins of error, usually on the order of plus or minus 0.25 degrees Fahrenheit. You are unlikely to see those margins of error mentioned in media accounts. The temperatures aren't posted until three weeks or so into the following month because reports from some countries are late arriving, and all this requires considerable analysis. The results might not be perfect, but Arndt points out that NCEI's numbers track well with other global databases, including those maintained by NASA's Goddard Center, Japan, and the United Kingdom.[9]

One thing is quite evident with the quickest scan of NCEI's monthly and annual reports: Warming is not uniform. The Arctic is a planetary hot spot, where the rate of warming appears to be about double that of the rest of the planet. Dave Robinson sees the evidence of that warming in the satellite data on snow and ice cover. Warming begets warming. Snow and ice are powerful repellents of solar energy, so when they get out of the way, the sun can go to town. In 2012, Arctic sea-ice extent slid to its lowest level since at least 1979, when the government began keeping score.[10]

What difference would that make to the people who live far closer to the Tropic of Cancer than the Arctic Circle? In terms of the future of snow in the United States, perhaps everything.

A World of Difference

We do care about the South Pole but in the United States not quite as much. That has something to do with American-centricity; that would be only in part. Despite the potential for massive ice breakup and calving, for now temperatures aren't increasing at the rate they are at the opposite end of the earth. It stands to reason that the Arctic would have a far greater effect on the midlatitudes than the Antarctic. And in terms of sensible weather, the Southern Hemisphere and Northern Hemisphere are very different from one another and not just in the realm of their mirror-opposite seasons and the circulation of cyclones; below the equator, winds blow clockwise around storm centers.

The Southern Hemisphere cool season, from fall to spring equinoxes, is a little less than five days longer than its northern counterpart, and the warm seasons 2.25 days shorter. February is the shortest month of the year because, as hard as it might be for North Americans to imagine, the earth makes its closest approach to the sun during the Northern Hemisphere winter, causing a speed-up in the planet's elliptical orbit. Solar angles trump proximity to the sun for the distributions of warm and cold. For a fascinating take on how this orbital quirk came to be and how it could reverse course in the far,

far distant future, I would highly recommend John and Kathleen Imbrie's readable *Ice Ages*.

In addition to the lengths of the seasons, a major difference between the two hemispheres is the amount of water. Given that the Antarctic snow and ice cover isn't as volatile as the Arctic's, and the fact that so much of the surface is covered by water, the government doesn't bother to monitor Southern Hemisphere snow cover.[11]

The ratio of ocean to land in the Northern Hemisphere is double that of the Southern. Temperatures change ponderously in the oceans, which absorb solar energy and redistribute it in powerful currents like the Gulf Stream. Temperatures on land can be jumpier than a bored teenager. Land surfaces heat irregularly since they can have radically different reflectivity: Forested areas do not absorb and retain solar energy as efficiently as prairie land. The west-to-east air currents that enter North America from the Pacific have to negotiate an obstacle course before they get to the Atlantic, ramming into mountains that deflect and distort the flow, cascading into the Plains, spinning up cyclones, and ramming into more mountains. As the Japanese learned from their World War II balloon-bombing campaign, the jet stream doesn't have a linear bone in its body. What they thought would be a straight shot from Japan to the U.S. Northwest forests turned out to be an often-serpentine journey that blew their balloons from Michigan to Texas and sabotaged their sabotage campaign. The jet-stream winds that bedeviled Allied pilots in World War II were invaluable but unpredictable allies during the balloon-bombing campaign.

Those jet-stream winds have been a focus of research for a reason: So goes those winds, so goes the future of weather in the United States. The winds form at the upper-air boundaries of warm and cold air, the mighty battlegrounds of tropical and polar air masses, as the planet seeks to maintain a reasonably constant temperature. Temperature contrasts drive winds and storms. That's why the jet stream, an esoteric concept until the mid-20th century, has become a media star. On The Weather Channel and TV weather reports you will see it represented in a deceptively cartoonish fashion, a fleshy colorful arc typically with a pronounced buckle sagging into the northern and/or central part of the nation in winter. If you don't like cold, you don't want to be on the north side of the buckle, and if you don't like winter storms, you don't want to be near the area where the dip reaches its southernmost point and bends back toward the north. That's often the area were the jet-stream winds spin up storms, lifting the tops off of cyclones, rocketing their upward rising currents high into the atmosphere's precipitation factories. Knowing the position and strength of the jet-stream winds is absolutely crucial to forecasting winter storms. When a jet stream becomes locked in place, prolonged cold snaps ice the areas to the north, while prolonged mild spells rule to the south.

Since temperature contrasts drive the jet winds, a warmer Arctic would mean retreating jet streams, which intuitively would argue for a similar retreat of snow lines and progressively less severe winters. The atmosphere, however, has long exhibited a healthy taste for the counterintuitive.

An Abrupt Scare

Some researchers have speculated that an utterly opposite impact would occur with the slowing of the jet stream's oceanic cousin, the Gulf Stream current. The two streams do share common traits. The atmosphere behaves like a fluid, and Gustav Rossby said that his research on the Gulf Stream, with its embedded "jets," inspired his interest in the jet stream.[12] The Gulf Stream is a massive heat exchanger that imports vast quantities of tropical and subtropical heat to the Arctic. Like the jet, it is not a simple "stream." Its offshoots, and tributaries contribute to moderating winters in the British Isles and parts of Western Europe.

We have mentioned *The Day After Tomorrow* only reluctantly; Hollywood has done enough mangling of the images of snow and ice. In the movie, the most incredible winter siege in the history of moviemaking entombs the nation and the rest of the world. This is not quite what Wally Broecker had in mind.

Broecker found evidence of a rapid climate change millennia ago in the presence of an Arctic shrub, known as the Younger Dryas. It appeared in Europe at latitudes far below where it would be present today. In a seminal 1987 article in the journal *Natural History*, Broecker described the oceanic circulation as a "conveyor belt" that transported warm water northward and sinking cold water southward. As everything else involving physical oceanography, the processes are immensely complex and not wholly understood. Cold, saltier water is heavier than the less salty, fresher water it encounters in the far North Atlantic, where glaciers have been melting for thousands of years. The heavier water sinks to the bottom of the ocean and flows back toward the equator. The overlying warm water flows back toward the Arctic. What if, for some reason, the Atlantic conveyor belt shut off? Could glaciers advance and lock much of the Northern Hemisphere in a reign of winter? Could that explain the equator-ward invasion of the Younger Dryas. "The Younger Dryas may be God's message to us that the climate system is more complicated than we believe," Broecker told me in an interview after the article was published.[13] Broecker's essay was followed by a wave of papers and discussions about the concept of "abrupt climate change." With so much attention having been drawn to his article, Broecker wrote follow-up pieces insisting he was making no speculations about the future and that his *Natural History* article had contained nothing predictive. It was about something that had happened.[14]

The 1987 paper did contain an illustration of the conveyor belt circulation that probably has gained far more attention and become more well remembered than the content of his article. The graphic was far more simplistic than reality. Two decades later, when I interviewed him at Columbia, he reminisced about that graphic and recalled that it essentially was a cartoon.[15] In real life, the ocean, like the atmosphere, is a chaotic place full of whorls and eddies. Unquestionably, it is both a driver of climate and perhaps its most important passenger, and any major changes in circulation would affect land masses profoundly.

Wally Broecker's conveyor belt concept had legs, arms, and wheels at the turn of the new century, and the possibility that the Gulf Stream could slow down or stop became a popular and chilling source of research and popular articles. The Woods Hole Oceanographic Institution, located on the southern Massachusetts coast, where oceanographer Ruth Curry worked, became an important center of that research.

When I met her at Woods Hole in 2005, what struck me was how she didn't fit the profile of someone who had just informed the world that the climate system might be in serious jeopardy. At the time she was 50, and she looked about half that. In a voice evocative of an announcer on a classical-music station, Curry rather clinically explained what she was seeing and why the world should be worried. Her specialty was freshwater intrusion into the North Atlantic from glacial melt. The water was freshening alarmingly, she said, and the sinking rate of sea-surface waters in the high latitudes might slow dangerously, in turn slowing the conveyor belt.[16]

Curry's boss had raised the profile of that message with an article in the Woods Hole magazine. It was well-written and reasoned and articulated, and while reading it one would not have been surprised to hear the *Jaws* soundtrack in the background. Talk of climate change traditionally centered on the concept of a "gradual increase in global temperatures," wrote Robert Gagosian, who then was the institute's director. That attitude is passé, he warned. "It ignores recent and rapidly advancing evidence that Earth's climate repeatedly has shifted abruptly and dramatically in the past. . . . Thus, world leaders may be planning for climate scenarios of global warming that are opposite to what might actually occur."[17]

He referred to computer models showing that the North Atlantic region would cool by 5 to 10 degrees Fahrenheit if the conveyor belt went on shutdown, and that "would produce winters twice as cold as the worst winters on record in the Eastern United States." Furthermore, "Severe winters in the North Atlantic region would likely persist for decades to centuries."

When the head of a venerable oceanographic research center makes such dramatic statements, it's bound to get media attention. It got mine, and I wrote a front-page story about it. Gagosian evidently got the attention of the U.S. Defense Department. Ten months after it was published, a Pentagon-commissioned report ominously postulated that slowly rising temperatures could set off rapidly falling ones. "Recent research . . . suggests that there is a possibility that this gradual global warming could lead to a relatively abrupt slowing of the ocean's thermohaline conveyor, which could lead to harsher winter weather." It was a familiar hypothesis: It cited Woods Hole's research.[18]

"It is quite plausible that within a decade the evidence of an imminent abrupt climate shift may become clear and reliable," the Pentagon authors wrote.

We in the media are apt to report dramatic statements emanating from the scientific community, just about every one of which has a counterclaim out there somewhere, and in my experience the rebuttals far outweigh outright refutations. The government has objective and subjective methods of verifying the quality of shorter-term forecasts. But what about forecasts on the decadal scale? Thirty, 50, 100 years from now who is going to remember who said what? In the 15 years since the Pentagon paper appeared, that evidence of "imminent abrupt climate shift" fortunately is still very much wanting.

Rather than resolution, from my observation the most pointed controversies lose their edges through sublimation, the way winds can erase ice and snow gradually. That has been the case with abrupt climate change. Some scientists still talk of the possibility, but it is not the buzz phrase it was at the turn of the millennium. When I spoke with him, Broecker said rather diplomatically that Gagosian had gone "too far."

I learned from talking with other oceanographers, for example, Ray Schmitt (now deceased) at Woods Hole and Bill Johns at the University of Miami's Rosenstiel School, and by spending time on research vessels, that the oceans rival the atmosphere for not only complexity, but also observation gaps. The Gulf Stream, for example, is not simply a surface current, nor is it a linear one. The width varies throughout time, as do its eastern and western boundaries, as does the depth of the warmth. To determine whether the flow is slowing requires monitoring the speed of the current. Given the size and magnitude of the current, is that possible?

Bill Johns and a team of researchers are trying. Using a deepwater instrument array that spans the Atlantic near the latitude of Miami, they are measuring changes in the oceanic circulation at depths from about 4,000 to 15,000 feet. What they've found is a reduction in the current and what appears to be a northward shift of the Gulf Stream system that is consistent with what global models predict in a warming planet. The caveat, however, is a significant one: The instruments have been in place only since 2004, and it is too soon to say how much of that behavior is due to greenhouse warming or natural variability.[19]

The career of the Gulf Stream is of importance not just to the East Coast. Should a slowing actually result in a significant cooldown in the East, that would affect weather worldwide. Nothing in the atmosphere happens in a vacuum. If that storm depicted in *The Day After Tomorrow* actually occurred, other parts of the planet would be on fire from the heat.

When a hurricane or winter storm threatens a given area, the public naturally focuses its attention on the path of the storm itself. Meteorologists certainly do also, but they also are well aware that a storm doesn't act on its own, but interacts with other systems on its borders—systems that in turn interact with others. In an abrupt-cooling scenario, the temperature drops would be regional while theoretically the global temperature continued to increase, which would suggest some parts of the planet would become dramatically warmer. Hypothetically, an East Coast cooldown could have a mirror-opposite impact in the West.

The fate of the oceans is crucial to the fate of the world's climate. Gagosian might well have been guilty of going "too far," in Broecker's words, which is a shame. Rereading his essay, he did make a tidy and eloquent case, wholly accessible to the lay reader, for the importance of the oceans and their coupling relationship with the air that overlies them.

Pacific Power

"The oceans and the atmosphere constitute *intertwined* components of Earth's climate system," Gagosian wrote.[20] As he pointed out, one of the most obvious examples is the periodic El Nino phenomenon in the tropical Pacific. During El Nino, which often ripens at about Christmastime and thus draws its name from the Christ child, surface temperatures over a continent-size expanse of the Pacific, 4 million square miles or more, become anomalously warm. Peruvians are well acquainted with it because the warm waters effectively erase their annual anchovy catch. As Bob Livezey, a longtime resident brain at the government's Climate Prediction Center (CPC), often repeated to me, the tropical Pacific is the engine that drives the atmosphere. In El Nino, the heating of the overlying air sets off explosive thunderstorm convection in the high atmosphere that distorts the jet-stream winds, which are North America's storm highways.

The power of El Nino was dramatically evident in the winter of 1997–1998, when virtually the entire atmosphere over North America was flooded with warm air. The winter was chary with snow in the East; less than an inch was measured in Philadelphia, despite a sequence of coastal cyclones that ripped sand off of beaches. The most memorable storm of that winter in the usually snowy reaches of upstate New York, northern Vermont, and southeastern Canada did not involve snow. It was heavy rain. Unfortunately, it was freezing rain. While surface temperatures remained below freezing, the El Nino–warmed upper air wasn't cold enough for snow.

A forecast for an accumulation of 0.5 inches of ice is sufficient to trigger an ice-storm warning in the Champlain Valley. As many as four inches of ice accrued in the January 1998 storm. The weight of the ice collapsed electrical towers and hydro pylons. Millions lost power, including half the population of Quebec. The storm was blamed for at least 35 deaths and caused an estimated $4.5 billion in damages. The destruction of millions of trees set back the maple-sugaring and apple industries for years.[21]

The talk of abrupt change centered on the Atlantic rather than the Pacific. But the clear lesson of 1997–1998 was that the behavior of the Pacific is of monumental importance to snow in the United States. The overarching reasons are evident: The oceans are massive repositories of stored heat, and the Pacific is triple the size of the Atlantic. Plus, very simply, weather moves west to east. As meteorologists have learned from the history of winter storms, observation holes over the Pacific remain a tremendous handicap for computer modeling. Tony Gigi, a veteran National Weather Service meteorologist in the East Coast offices, has observed that he doesn't put much stake in snow forecasts until three days before the precipitation has started. That's when storms from the West enter North America, and their behavior can be captured by land-based instruments. The transcontinental trip typically takes about three days.

The future of the climate system obviously is inextricably tied to the future of the Pacific. In the global temperature data sets, the years coinciding with El Nino events rank among the warmest on record, a direct result of the super-warmed tropical Pacific. In 2018, Carnegie-Mellon University researchers found evidence of the Pacific's impact about 5,000 miles north of the El Nino zone. The warming waters of the far North Pacific are contributing to the rapid Arctic warming, they concluded.[22]

The changes in the Pacific, Atlantic, Gulf Stream, and jet stream likely will play out throughout decades as the planet's atmosphere continues on its merry, chaotic, nonlinear way. But the Arctic warming clearly has been happening in a hurry, and scientists in various meteorological and oceanographic disciplines are trying to unravel how the oceans and upper-air winds both drive and are driven by changes in the Arctic. We media types are shameless and ruthless in pursuit of word economies, and that's one reason why you so often will read and hear words like "climate researchers" and "scientists." It is reasonable to ask, "Just who are we talking about?" Shame on us for not addressing that question often enough.

Yet another important lesson I learned throughout the years through trial and error, particularly the latter, is that meteorology and oceanography are highly specialized disciplines. I once had an idealized view that "science" was a preserve that had a franchise on objective truths. I suspect that any practicing "scientist" who read the preceding sentence, if not convulsed with laughter, would agree that only a nonscientist could have held such a belief. In the weather community I am always surprised at how often meteorologists don't agree on what they see, let alone what's going to happen. Almost everyone is opinionated; smart people tend to be more so. How often have you heard, "Great minds think alike?" Much closer to the truth is the axiom, "Mediocre minds think alike."

Given the complexities and enigmas of the climate system, it is quite remarkable that so much agreement has emerged among the diverse disciplines that Earth has warmed and will continue to do so. Some of the consequences are indisputable.

Sea levels have been rising, not linearly, but rising.

Summer nights are getting warmer, particularly in urbanized areas. Hot nights constitute a deadly hazard in the nation's large cities, something my newspaper has reported on extensively, and I'd like to believe that our reporting has made a difference in reducing heat-related mortality. Philadelphia has experienced deadly summers, and the Centers for Disease Control has commended the city's work in educating the public about the hazards and taking actions to prevent fatalities. Heat is particularly dangerous in row house neighborhoods that have high elderly live-alone populations. Eric Klinenberg's *Heat Wave: A Social Autopsy of Disaster in Chicago*, which chronicles the 1995 heat wave blamed for killing more than 500 people, is an ultimate study in urban heat hazards.

Growing seasons are getting longer, winters shorter, as the distance between last-freeze and first-freeze dates has been shrinking and is likely to continue doing so.[23]

Global Warming and Extreme Times

Harder and more specific questions remain a source of fierce discussion and down-shouting: How much and how quickly will warming accelerate? What precisely can be done about it? How will enhanced warming affect sensible weather?

Two of the more heated battlegrounds involve the future of the hurricane and winter seasons, and both have been tied to the future of the oceanic conveyor belt.

I have reported and written extensively about hurricanes in the last 20 years. Tropical storms are not only intrinsically fascinating, but also of immense social consequence. My Pulitzer Prize–winning former colleague, Gil Gaul, and I wrote a series published in the *Inquirer* about coastal overdevelopment and its consequences. Gil pursued the subject in his *The Geography of Risk*, which was published in September 2019, and which I can't recommend highly enough. We found that in terms of persistent hazards, disaster dollars, and stress on the U.S. Treasury, nothing rivals hurricanes.[24] No, tornadoes aren't even close, not even earthquakes. In the course of reporting that series and in the years after it was published, I have had the opportunity to meet and interview some of the nation's most prominent and knowledgeable tropical-cyclone experts. In the process I have been impressed with the levels of specialization in tropical-storm research and forecasting.

One of my favorite hurricane figures was the late Bill Gray, who would become a public enemy among "environmentalists"—again, one of those newspeak terms that is an injustice to an entire spectrum of specialties and scientific engagement—for his bashing of climate change. Dr. Gray developed groundbreaking methods of forecasting hurricane activity months in advance. Those methods have been copied by private and government forecasters, and still form the bases of all those extended outlooks. He was generous with his time, his expertise, and, assuredly, his opinions.

I visited him at his office at Colorado State University in Fort Collins, long one of the nation's most important meteorological research centers. Gray was an East Coast defector whose weather interests were forever vectored by his experiences with Hurricane Hazel in October 1954. Forty-five years later, Gray should have been basking in triumph. He had identified "lull" and "active" hurricane periods in the Atlantic Basin, each lasting 25 to 40 years. He had warned coastal residents and property owners along the U.S. coast that the building boom in the '70s, '80s, and early '90s would come at a price. That construction had occurred during one of those lull periods that was sure to end. He was correct. In 1995, the basin entered a period of unprecedented activity in the period of satellite observation.

Rather than celebrating his achievement, however, Gray was complaining that he couldn't get grant money to continue his research. He pulled out sheets of diagrams hand-drawn on yellow legal-pad paper. The penciled drawings looked like crosses among illustrations of molecules and old-fashioned vacuum cleaners. "I can't get funding for this," he said. Meanwhile, his younger colleagues were getting massive infusions of federal dollars for what he called "big science." Holding up his drawings, he declared that big science "can't compete with this." Unquestionably, Gray's work represented a breakthrough in long-range hurricane forecasting, and while he wasn't right every year, his forecasts far surpassed what would have resulted from applying mere climatology.[25]

One thing he insisted upon, however, was perplexing to my lay mind. He published four outlooks a year, one in December, followed by updates in April, June, and August. They were full of fascinating discourses on such factors as rainfall in the Sahel and sea-surface pressures over the Atlantic and the quasi-biennial oscillation. He said the active and lull periods in the Atlantic were tied to large-scale changes in the "thermohaline" oceanic circulation, changes in temperatures, salinity, and speeds of currents. Other researchers have agreed that hurricane activity is tied to a circulatory cycle known as the Atlantic Meridional Oscillation, or AMO, and yes, that, has been a source of debate.

What stopped me in Gray's outlooks was the statement that hurricane activity increases when the south-to-north thermohaline circulation speeds up, and that's what has been happening. He made the same assertion in a PBS documentary.

I'm always wary of those clean relationships, but in my unscientific mind, I would have thought the opposite would be true, that a slowing of the circulation would mean less heat transport from the tropics and subtropics, and thus higher sea-surface temperatures in the hurricane breeding zone, thus more fuel for tropical storms.

I asked Ray Schmitt at Woods Hole about Gray's hypothesis. His response was that Dr. Gray was well regarded in the research community for his pioneering work on hurricanes, not so much for his oceanography. Bill Johns, who was involved in that Herculean effort to measure the flows in the Atlantic, told me that he once asked Gray how he came to know that the Atlantic circulation was speeding up and what data he was using but didn't get an answer. Whatever Gray had to say about the oceanic circulation, or climate change for that matter, no one matched his achievements in identifying the macro forces that drive hurricane seasons. (His prodigy, Phil Klotzbach, has continued his excellent work on tropical storm forecasting.) Gray could well have been right about the speed-up of the Atlantic; the evidence is lacking for an abject refutation. But the lesson might well be that it is impossible for even the smartest people to know everything.

Unlike Gray, some other researchers heartily embrace the role of anthropogenic warming in hurricane development and intensity; the specifics are another matter. Michael Mann at Penn State and MIT's Kerry Emanuel, two prominent figures in the climate change community, believe the Atlantic Basin is locked into a permanent active cycle and are bullish on increasing intensities.[26] Frank Marks, longtime head of the government Hurricane Research Center, has been more circumspect. Marks has spent a career working on the hurricane problem and flown into the eyes of potent cyclones numerous times.

Even if worldwide warming resulted in warmer sea-surface temperatures in the hurricane-spawning zones of the Atlantic Basin, which includes the Gulf of Mexico and Caribbean Sea, Marks said that wouldn't necessarily mean stronger storms. Hurricanes require a matrix of circumstances to ripen. "You'll have to show me more than warm water," he said.[27]

After Superstorm Sandy caused catastrophic damages in the New York City area in 2012, the governor and mayor decried what they viewed as the unmistakable consequences of climate change. The average sea level in New York Harbor has risen about a foot since 1900, as a result of thermal expansion, according to the city's Department of Environmental Conservation,[28] and that higher base water level certainly was a flooding additive as the storm made landfall near Atlantic City, New Jersey. But did global warming make Sandy more intense and cause a counterintuitive jog to the west as it approached the New Jersey coast? Marks opined that if Sandy had made landfall at low tide, no one would be calling it a superstorm. That moniker, by the way, not only applied to the storm's strength, but also spoke to the ambiguities surrounding whether it was properly a "hurricane" when it made landfall. Technically, the government declared it an "extra-tropical cyclone" since it had lost its tropical characteristics.

The affirmation that the planet is warmer and that the warming might accelerate has emptied into disturbing uncertainties.

The 2011 tornado season was so destructive and relentless that not surprisingly, questions arose about the influence of rising temperatures. In a single outbreak in late April, 122 tornadoes touched down in Tornado Alley, and 319 people were killed. That season I visited the Storm Prediction Center in Norman, Oklahoma. I watched the forecasters on a chaotic afternoon where so many tornado-alert bells were going off I thought I was inside an old-fashioned pinball arcade. When I saw an alert go up for Caribou, Maine, I figured I had seen everything.

Away from the chaos of the forecasting-operations center, I met with Harold Brooks, one of the nation's most knowledgeable sources on tornadoes. I'm not sure what it is about the culture of government meteorologists, but to a person from National Weather Service chief Louis W. Uccellini on down, they are among the most unaffected human beings I've encountered. Brooks was no exception. For a man who had been through one hellish season, he was remarkably calm.

He did not see the direct hand of climate change in the tornado harvest of 2011. For example, he said, it would not be possible to tease out the impacts of ultra-warm waters in the Gulf of Mexico. As hurricanes, he said, tornadoes require very specific conditions through different layers of the atmosphere. Those alignments would have to include more than abnormal pulses of Gulf moisture. The horrific casualty figures were tragic, but Brooks explained that some of the 2011 tornadoes defied overwhelming odds in that they targeted people and development in regions that were otherwise sparsely populated. The mathematical likelihood of a tornado path causing mass fatalities in those regions was almost infinitesimal, he said. Improved warning times no doubt have saved lives, but tornado tragedies continue to occur. For now, Brooks said, it would be impossible to determine precisely what global warming contributes to them.[29]

The Future of Snow

Ultimately Brooks and Marks worked for Louis Uccellini, the world authority on winter storms. At the time he was head of the government's storm mayhem centers. He later became the chief of the entire National Weather Service, and I can't imagine that job is a barrel of laughs. Uccellini has to make personnel decisions, attend meetings, and shovel mass quantities of paperwork. Still, he loves to talk about snowstorms. In 2016, I was invited to an "informal" media session at headquarters in Silver Spring, Maryland, where he was to brief a small group of journalists about changes coming to the agency. I had other duties at the paper and a reflexive journalistic reluctance to accept the invitation; I prefer to ask my questions privately rather than in the presence of other reporters. For one thing, I'd like to keep any good information to myself and my publication; for another, I would prefer to confine sounding stupid to one person rather than a group.

In the end, I decided to get over my hesitation. Uccellini's media person, Susan Buchanan, said the boss wanted me to come because he really was looking forward to chatting with me about "gravity waves." Those are atmospheric waves that can result in profoundly heavy snows along narrow corridors in potent cyclones, and they are described in the seminal two-volume set *Northeastern Storms*, cowritten by Uccellini and Paul Kocin, who authored the scientific treatise on the Blizzard of 1888. Gravity waves help explain how it happens that a given corridor can get two feet of snow, while areas 20 miles away on either side might get half that.

Uccellini has a long-standing interest in gravity waves; they were the subject of his master's thesis in 1972. In their book, Uccellini and Kocin cited the February 11, 1983, blizzard as a classic case study of the impact of a gravity wave rippling along the I-95 corridor, setting off profoundly heavy snows, with lightning and thunder. I well remember the storm, and I happened to be walking in Center City Philadelphia at about 4 p.m. I had never experienced snow that heavy. As it turned out, according to Uccellini and Kocin, the wave was approaching Philadelphia at about that time.[30] It proceeded to New York, where five inches of snowfall was measured in just an hour, trapping motorists inside the Lincoln Tunnel. If you have spent any time in that tunnel, you likely agree that it would be an experience you would happily live without.

After that "informal" meeting, which turned out to be quite interesting and useful for string-gathering, Uccellini summoned me to his office to talk gravity waves and share an article he found particularly interesting.

He clearly was more comfortable talking about gravity waves than the uncertainties of the effects of climate change, but the subject no longer can be avoided. During the actual meeting, Uccellini said he does strongly suspect that it is juicing up winter storms, but he stopped short of saying the case for a greenhouse signal in winter storms was 100 percent sealed. As he had in the past, he did refer to the "enhanced precipitation signal."

In a 2013 interview posted on the University of Wisconsin Space, Science, and Engineering Center site, Uccellini discussed what global warming might have to do with snow. Wisconsin hosts one of the nation's elite meteorological programs. In the late 1960s, when Uccellini first showed up in Madison, the Wisconsin campus was a hotbed of radicalism and wide-eyed political idealism, but even back then Uccellini apparently stuck with the science. He completed his undergraduate, graduate, and doctoral work there, and joined a distinguished pantheon of alumni. He holds a position of responsibility and power, his expertise is admired even in the private sector, and it doesn't hurt to have a meteorological glossary when you talk to him. But he has a gift for putting people at ease that has served him well, and the science-speak vocabulary aside, chatting with him I felt I could well be talking to someone on the Corner in the working-class neighborhood of my youth. He grew up on Long Island and still has traces of the venue in his voice. Its volatile weather environment, notably the effects of hurricanes Carol, Diane, and, particularly, Donna, in 1960,[31] fired his interest in the atmosphere.

As for what climate change has to do with winter storms, intuitively it makes sense that a warmer atmosphere would hold more water vapor, which would lead to more condensation. When water condenses, it surrenders the latent heat that it stored when it executed its magic act by vanishing into vapor. The released heat further powers storms, which leads to more latent-heat release.

"There is evidence to suggest that this increase in warming and related water vapor is playing a role in more extreme storms," he said. "We're seeing more cases of extreme events and heavier precipitation."[32]

In the longer term, increased atmospheric warming and changes in the oceanic circulation could well conspire to erode the world's snow cover. But the atmosphere appears to have that insatiable appetite for the counterintuitive, something that is evident in Dave Robinson's snow maps. In the short term, warming might be enhancing snowfall in parts of the United States. Robinson serves on a committee that is establishing what literally will be the "new normals." Those normal values are recalculated each decade based on the atmosphere's behavior in the previous three decades. The 2011–2020 normals were based on data from 1981–2010; the 1991–2020 data is the basis for the 2021–2030 normals. Philadelphia likely will be an example of the coming paradox, he said. The temperature values should be slightly higher, but the snow averages also will be slightly higher. In the 21st century, several snow records have been buried from Washington to Boston.

The "enhanced" precipitation phenomenon referenced by Uccellini would explain the increased moisture supply attendant to that recent run of mega-snowstorms. But storms need cold air, and Northern Hemisphere land masses have been warming more robustly than the surfaces of the rest of the planet. That is evident in the government's monthly and annual global temperature reports.

What explains the paradox? A group of scientists led by Judah L. Cohen, who has had some success at correlating Eurasian snow cover in October to the North American winter, took on that question in a 2012 paper. Yes, Northern Hemisphere land temperatures have warmed more than the globe, but they found that the temperature increases in the warm seasons have outpaced those of winters. That would correlate with Robinson's finding that wintertime snow cover generally has held its own but has been disappearing in a hurry in the spring months. Cohen and his coauthors argued that the diverging seasonal trends could be very much related. In the previous two decades, they wrote, a winter cooling trend has been the rule in parts of Northern Eurasia and eastern North America. They reasoned that the cooling wasn't the result of "natural variability," a catchphrase, and a perfectly suitable one, for the unexplained. Rather, they believe that the winter trend was associated with generally increased levels of atmospheric moisture in the higher latitudes—thus more cloudiness and precipitation—and that could have been the result of higher water-vapor content produced by the higher temperatures in the warm seasons.[33]

Furthermore, more water vapor has led to increases in Eurasian snow cover, which is an effective solar radiation repellent. Cohen has postulated that Eurasian snow cover in the fall is a driver of a crucial atmospheric phenomenon known as the polar vortex. That poor thing was able to spin in obscurity for millennia and then rather suddenly became a superstar when the media discovered it during the winter of 2013–2014. (Just for the hay of it I tried a search on the term "polar vortex" for all the newspapers in the Nexis system for the entire period predating that winter and found exactly one match: It was a reference in a 1984 *Miami Herald* article about sharks by former colleague Mike Capuzzo.) It is a menacing term—"It sounds like a horror movie," said Steven Strader, an atmospheric scientist at Villanova University—for a quite natural phenomena.[34] Glenn Schwartz, a long-tenured TV meteorologist in the Philadelphia market, said that when he was getting into the business, he was warned never to use such a technical term on the air. The polar vortex, as its name implies, is a powerful mass of air circulating counterclockwise in the high atmosphere above the poles. It intensifies during the long night of winter. It has been likened to an immense bowl of frigid air spinning madly at the top of the world.[35] On occasion, the rotation weakens, and some of that hyper-chilled air sloshes out of the bowl and spills southward, and temperatures plunge in parts of the United States. In simplistic terms, the U.S. winter is governed by how much cold air can spill out of the Arctic.

The behavior of the polar vortex is a crucial component of the air-mass transfer. It affects the behavior of air masses that are closer to the surface and at lower latitudes. That phenomenon is captured in the indexes of the Arctic Oscillation (AO), which is a prime mover of the winter across the hemisphere, and its close relation, the North Atlantic Oscillation, or NAO, which Louis Uccellini said is pretty much the ballgame for the winter in the Eastern United States.[36] The fates of the jet-stream winds, those battle-grounds between cold and warm air that are the igniters of storms and the bringers of winter, are bonded inextricably to the polar vortex and those indexes.[37]

To monitor and predict winter weather in the Eastern United States and Western Europe, the NAO is tracked by an index that measures pressure differences over the ocean between the latitudes of Greenland and the Iberian Peninsula. When pressures are lower at the higher latitudes, likely the case when the polar vortex is strong, the index is said to be positive. That has the effect of repelling cold air from the East. When the pressures are higher at the more northerly latitudes, the index is positive, and that refrigerated air can penetrate southward.

Two winters a decade apart—2009–2010 and 2019–2020—make for a convincing case. The Mid-Atlantic and Northeast were smothered in unprecedented snows in 2009–2010, when the NAO was more negative than a political commercial. Conversely, in a season when the polar vortex was consistently intense, the NAO index stayed outrageously positive in January and February 2020, which were therapy-dog situations for jilted snow-lovers in the Northeast. "They're hurting puppies," said Uccellini, himself a snow-lover.[38] (In late January, one chat board started a "Winter 2019/20 Melt/ Implode Thread." Said the moderator, "My time is coming, my feet are on the ledge.")[39]

The NAO's influence on winter is clear. For longer-range forecasting, the challenge will be to figure out precisely what drives the NAO. The stratosphere appears to the NAO's puppeteer, said Uccellini. That's a problem: That layer of the atmosphere is miles and miles deep, and the data is sparse. Another issue: Uccellini points out that the NAO is not even an oscillation per se; it changes in a hurry. Unlike the El Nino/SO, which is driven by slow-changing ocean temperatures, the NAO is a child of the whims of an atmosphere that often lacks an attention span. For now, the course of the index remains unpredictable beyond five to seven days.

Judah Cohen, who works for the Atmospheric and Environmental Research firm, based in Massachusetts, believes he has found one promising clue for predicting the course of the NAO and the winter. His research has found links between the autumn Siberian snow cover and a weaker polar vortex in the subsequent winter. Thus, more Siberian snow could mean a colder and wetter U.S. winter. Uccellini and Mike Halpert, who has been the lead seasonal forecaster at the government's CPC, says the correlation is promising, but they also note that in some seasons it just hasn't worked, something Cohen readily acknowledges. But he says his forecast whiffs have value: "You learn more from your failures than your successes."

Although a clear winter-phile, Cohen even saw a positive in suffering through southern New England's disconcertingly mild and snow-scarce months of January and February 2020. Since the powerful and relentless polar vortex was confined primarily to the Arctic, it was a decent winter for rebuilding the rapidly depleting polar sea ice.[40]

More researchers have been using "North Annular Mode" in the literature to describe both the NAO and AO, said David W. J. Thompson, an atmospheric scientist at Colorado State University. Throughout most of the 20th century, he said, they referred to the NAO almost exclusively because both the available data and the strength of the anomalies are the largest over the Atlantic. The AO was added late in the century to denote that the pressure pattern did cover the entire hemisphere. They both constitute the Northern Annular Mode (NAM). The *annular* refers to the fact that the overall oscillating pattern is ring-like.[41]

The connection between the career of the Arctic and midlatitude snows has been a hot source of speculation. A weakening polar vortex, which would promote that warming, could make the jet-stream winds "wavier" and thus more prone to buckling and plunging deeply into the heart of the midlatitudes. That was the premise of a paper by Rutgers University researcher Jennifer Francis and associate Stephen J. Vavrus.[42] A weaker vortex would promote more of a north–south flow—that is, meridional or "high amplitude"—in the upper atmosphere rather than west to east, or "zonal." That would make for more extremes of temperatures and precipitation.

"The frequency of days with high-amplitude jet-stream configurations has increased during recent years," wrote Francis and Vavrus. "Notable examples of these types of events include cold, snowy winters in Eastern North America during winters of 2009–2010, 2010–2011, and 2013–2014; record-breaking snowfalls in Japan and southeastern Alaska during winter 2011–2012; and Middle East floods in winter 2012–2013, to name only a few."[43]

Francis's conclusions have taken some heat in the scientific community, and although the hypothesis remains a source of debate, it does have supporters, including Rutgers colleague Dave Robinson.

No one in the research community or on the front lines of forecasting has found the Rosetta Stone to interpreting the atmosphere's chaotic signals, and the people involved in seasonal forecasting have had long relationships with frustration. For foreseeing the weather months in advance, the dynamical computer models just aren't there yet. They keep trying, and every year it seems the winter outlooks—and these are far and away the ones that capture the most interest from the pubic—are issued earlier and earlier, some of them in summer. The ones posted by the private sector, including TV stations, take their shots at estimating snow, crawling out on the most tenuous of limbs. Snow is hard enough to nail 12 hours in advance, sometimes two hours, and at times when it is still falling, let alone six months. Snowfall is the result of specific encounters of moisture and cold air, and it is quite possible that a cold and wet winter in any given area could end up being relatively snowless. As Uccellini says, snow isn't "periodic," it is "episodic."[44]

Snow Chance

Having wrestled with the madness of the atmosphere throughout his career, Mike Halpert has bonded with human fallibility. The climate center's seasonal outlooks are far sparser than those produced by such commercial companies as AccuWeather and The Weather Company, which are decidedly more attention-getting. CPCs are confined to mapping probabilities for above- and below-normal precipitation for vast regions of the country. They are ambiguous for the simple reason that they stay within the limits of the science. CPCs eschew snow and precise temperature forecasts. The "prognostic discussions" that accompany the monthly and seasonal outlooks are full of great information about "ENSO" and "canonical correlation analysis," for those who care to read them; however, in the longer-term forecasts, typically the most effective "tool" is the "optimal climate normal," a fancy term for the trend. And the trend has decidedly favored milder winters in much of the United States; again, land-mass warming has exceeded that of the globe in general.

Said Cohen, "In this climate-change environment it's been proving hard to get sustained cold."[45]

Halpert has been cautious on the subject of how climate change is affecting winter, and at a winter-forecast briefing in October 2018, he was circumspect in addressing a question about global warming. When I interviewed him later, he again was guarded.

"Climate change is always lurking in the background," Halpert told me in that conversation. It might well be behind shifting precipitation patterns or contributing to the persistence of certain upper-air patterns. That said, "The extremes are part of natural variability." There's that phrase again. He suggested that if the atmosphere kept a diary, it would read like it was written by a madman. Embedded in the long-term trends are decadal cycles. The 1960s in the United States were seriously cold; the 1930s, warm. But not all the 1960s winters were cold and snowy, while some of the '30s winters were eventful; in Philadelphia the 11 below zero Fahrenheit reading on February 9, 1934, has never been matched. "There is so much randomness," Halpert said, a statement that no ski operators would dispute.

"I'm sure that the American people, and the world, are destroying the planet," said Blue Mountain's Bob Taylor, but "it still gets cold, it still gets warm, it rains, it snows."[46]

"The planet is warming. That's indisputable," Halpert said in a later conversation.[47] That's one reason the climate center's seasonal outlooks have tended to favor above-average temperatures.

Opting for warmer in a winter forecast generally is a safe bet, as it was for the winter of 2019–2020; however, Halpert said that if you go with it every winter, "some years you're going to be woefully wrong."

The day might well come when it will be too warm to snow in New York or Boston or Chicago or Denver, throughout the country, throughout the world, but that day isn't imminent. In the short term, if anything, the additional water vapor in the air resulting from worldwide warming could lead to more mega-storms. "One could hypothesize larger snowstorms when all the stars align," said Robinson. Maybe the smaller snowstorms will be less frequent.

"It will continue snowing."[48]

The world without snow would be an immeasurably impoverished place, even for those who have suffered its inconveniences and dread the mere mention of the word in a forecast. They might well be nostalgic for those inconveniences; they might be surprised at what they end up missing. For the millions of its admirers, those who hold that rain is of the earth but snow is of the cosmos, it would be beyond the unimaginable.

"The pattern of a snowflake is a miniature of the design of the whole universe," wrote Caryll Houselander. "Snow is an enchantment laid upon our lives, woven into the lore of childhood, the image of the human race."[49]

EPILOGUE

Panic and the Pursuit of Happiness

Behind the sagging cyclone gates, the fast-falling snow awakened the forgotten details of the imprisoned copper pipes and the forsaken carburetors and refrigerator motors that Bill Morgan had purchased from his customers. With his cumbersome work gloves Bill affixed a padlock to the frigid, fat-linked chain he had threaded through openings where the sagging gates joined. It was early for a Saturday, but he was going home.

Bill owned the junkyard. It was Chester's junk room, the way station for the artifacts and residue of a fading river town, recycling before there was recycling. He bought the business from a patrician family that had moved out of town. We would bring him our papers and rags in exchange for nickels and dimes. (And do I ever sound like my father telling stories of *his* youth.) He pretended to weigh the bounty diligently on his crude scale to arrive at a fair settlement, scratching the figures in pencil on white note paper; it wasn't about the money. Maybe it was for the grown-ups, who sold their scrap carburetors and copper piping, not for us. While he weighed, we talked, about baseball, about anything. Bill was the only black man in the neighborhood, and thus the only black adult with whom I had a relationship. Each year before we went on vacation, he would give us a box of comic books that no doubt once rested on his scale and for which he had paid the appropriate settlement. How grateful I was to find him here in this transformed landscape, the only human being in the outside world doing business on this white and stormy evening. The talk that evening was about snow, and it was ever so brief. He didn't have much time.

Saturdays were Bill's rush hours. Kids in the morning, metal-hawkers in the afternoon. He might hang around until 9, imposing order on the metal scraps or organizing those white note papers with his random pencil marks, but not tonight. He was packing up, heading home to take shelter from the storm. A nine-year-old couldn't miss the message: I would be wise to do the same. I would be all alone now to experience the first blizzard of my life. I turned toward the wind to head home, my exhilaration mixing with a cold, creeping sensation of fright as the heavily falling snow erased the familiar and imposed a new order.

Milk, Bread, Eggs, Panic

Had that storm occurred today, by the end of the workweek the supermarkets in the Mid-Atlantic and Northeast would have undergone a sacking worthy of the Visigoths. So much has changed. In my youth, we didn't know about snowstorms until they were imminent, if then. We would hear about them the way we heard about World Series scores—via what one writer called the "ribbon of rumor"—when the games were played on weekday afternoons.

That was long before the era of supermarket snow panic, that altogether peculiar phenomenon that is at once mystifying, seemingly irrational, and understandable. It is part byproduct of one of the scientific community's fantastic success stories.

I've read and admittedly written tiresome stories about the irrational aspects. How many times can reporters and the cameras go to grocery stores to chronicle that the shelves have been raped of bread, milk, and eggs. It is indeed true that those are three immensely popular and scarce items when snow is in the forecast, and the French toast references are inevitable.

That behavior justifiably has been a source of mockery, likely by some of the same people who were out there buying the bread, milk, and eggs. It has also been defended.

I would argue that in our efforts to report and explain this behavior, we overlook the poetical, something that transcends our reasoning.

To make that argument, it would make sense to start at the beginning, except no one is quite sure where that might be. My mother fed nine people, and I never knew her to run out to a store when snow was in the forecast or feel any need to stock up. In my memory, we always ate heartily when it snowed.

I've come across different theories of when this all started, but in the Philadelphia region, which is a panic nexus, I don't recall reports of super-market stampedes before the very well-forecasted 1978 and 1983 blizzards.

Going through our archives, the first story I could find that talked about presnow grocery lines appeared in January 1988, before a snowfall that didn't turn out to be much of a deal. The subsequent four winters happened to be snow-deprived in the region, but then came March 1993.

"I see five gallons of milk in one cart alone," said a store manager in Lancaster County, Pennsylvania, where horses and buggies still share the roads with cars and SUVs, ordinarily a tamer part of the state away from the populated urban corridors. At 9:20 a.m. on Friday, the eve of the storm, lines had formed at nine checkout counters; usually the store had only four open that time of day. "I think people, to tell the truth, just, like panic," said a K-Mart manager.[1] Similar references echoed in newspapers elsewhere along the storm's path, even in Vermont.[2]

Like salting the roads, panic shopping officially had become an essential component of the prestorm ritual. In the 21st century, the supermarkets haven't looked back.

Before what turned out to be a truly historic snow siege in the Mid-Atlantic in 2010, I visited a Wegman's in Chester County, 25 miles west of Philadelphia, to watch the rodeo. "Insane is an understatement," the store manager, Kevin Russell, told me. He had just put in a 15-hour day. "Insane in a good way." The meat refrigerator was plucked clean, save for a discarded bag of baby spinach. Bagels were disappearing as fast as they were removed from the oven. By 9 a.m. on a Thursday, 36 hours before the first flakes, he knew he was in for a run on the cash registers. By 11, he was ordering managers to man the registers. He deputized his chef to be a bagger. The market expected a rush, he told me, but this was off the charts.[3]

A panic wave generated by a forecast swept across the Northeast four winters later with some customers alleging that bread and meat had been pilfered from their carts while they were looking away. The hoarding hordes responded to the snow-alarm bells again the following January, with shoppers sharing photographs of chaotic scenes inside supermarkets in Boston and New York.[4]

Wegman's insists that in the end, the snow stampedes have little effect on their revenues since they are counterbalanced by subsequent lulls. But that's not quite how it works, former Kroger's executive Jere Downs told me.[5] Kroger is by far the nation's largest chain, claiming better than 15 percent of the market, with more than 450,000 employees and reaping $121 billion in revenue in 2019.[6] It also has 37 manufacturing plants where it produces baked goods and dairy products,[7] the very things that tend to vanish from the shelves during the panic runs.

The mild, snowless winter of 2019–2020 was no friend of Kroger's, which has been contending with the growing threat of competition from Amazon. Said Downs, "In January and February grocers count on snow to drive customers back into stores, balancing out the dip in sales value following the holidays. This is really important. A good snow panic can drive the average sale from a weekday as high as a typical Saturday or Sunday."[8]

"Put another way, snow saves jobs."[9]

The hardware business is similarly snow-sensitive. A longtime Philadelphia-area store owner told me that his post-Christmas winter business is all about snow and cold, about selling the scrapers, shovels, and melters. The snowless January and February were positively brutal.[10]

But he also was aware that he could be taunting nature if he stashed away the snow-fighting products to make way for the rakes and grass seed.

One good storm threat could make for a lucrative panic run.

And it doesn't even have to snow.

What Has Changed

Five forces appear to be feeding the queues at the checkout lines when snow is in the forecast—science, the media (guilty), the actual weather, the motor vehicle, and an imponderable, human psychology.

Regarding the scientific component, those computers that attempt to solve the equations that predict the career of the atmosphere have redefined forecasting. As discussed in chapter 5, on almost any winter day, a computer model is seeing the potential for a major snowstorm or maybe more than one somewhere in the country in the next 10 days.

Some of them even happen.

The successes argue for taking the threats seriously. The model guidance does become more trustworthy as the predicted event approaches. If the bulk of the model runs are holding on to the snow within 48 hours of the scheduled arrival of precipitation, let the panic commence.

That the term "model" has gained such public acceptance is quite remarkable. I recall that during that 1983 weather conference in Washington, a science writer for a major newspaper complained that a presenter was using that word, and would he kindly explain what the hay he was talking about. I guess that was more or less like asking an architect what a blueprint is. He patiently explained that one couldn't very well build a scale model of the atmosphere and that computer simulations were a far more practical alternative.

These days, even if they have no interest whatsoever in how the models function, most people would accept that these technological marvels can generate forecasts on out to days in advance and see storms long before they exist. That acceptance has something to do with broadcast meteorologists who these days freely discuss and display what the "models" are showing.

That it is almost impossible not to know that a storm is coming has a whole lot to do with the likes of Joel Myers, John Coleman, and Frank Batten, as discussed in chapter 7. Joel Myers grew up in Philadelphia; he was into weather at a young age like Mozart was into music. Today, he would be a kid in a meteorological candy factory for which he was very much responsible. The boy who was jonesing for the weather became the man who dispensed it, the very person to whom that young Joel Myers would have loved to talk. He made his dream come true by creating and expanding AccuWeather. His company has provided around-the-clock weather to media outlets and newspapers throughout the country, and web users throughout the world.

With Batten's backing, Coleman established The Weather Channel (TWC), a once-preposterous concept whose presence would become ubiquitous. If a major storm is imminent, it's a safe bet somewhere near you, TWC is on a screen. If you're in an airport, hotel, or bar or restaurant with monitors, you won't be able to escape the news that something bad is heading your way. Complete with live images from your town or a snowy venue nowhere near you, TWC may be delivering its chilling prospects right in your living room; if not, your local TV station is—or your phone, which might even be beeping a storm warning. Better check the refrigerator and the battery supplies.

Forecast busts do happen, but so does snow, and for reasons that might well be related to a warming world, the late 19th century and new millennium have produced a harvest of mega-snowstorms, in some cases unprecedented in the period of record, and they have affected deliveries and people's abilities to get to stores.

In January 1996, when the storm tom-toms were beating for days, I steadfastly refused to yield to the hype and informed my wife I was not going to surrender to the region-wide panic. I did go to a pillaged market on the day before the storm to get two things that I would need whether it was raining, snowing, or partly cloudy: a pound of coffee and a six-pack of seltzer.

It was snowing heavily when I woke up at daybreak the next morning. Predictably, I got called into work and somehow survived the scariest 20-mile drive of my life. Naturally, I was stranded in the office, but at least my wife and our three- and seven-year-old sons had a pound of coffee, which only I would be drinking, and six 12-ounce cans of seltzer.

So, yes, not all panic shopping is irrational, and psychologist Jaime L. Kurtz has argued persuasively that it actually is quite rational and that the motivations go beyond risk assessment.

In the crosshairs of a storm threat, we need some sense of control. "We hate feeling helpless," Kurtz wrote. By loading up on foodstuffs, we can rest assured that "we did what we could."

And panic shopping feeds on itself. The bare shelves might move a consumer to ask, "What does everyone else know that I don't?" Furthermore, the fact that everyone is doing it creates a strange sense of community among the people who are participating in what has become a prestorm ritual.

Kurtz concluded, "This basic tendency to panic-shop actually makes perfect sense to me."[11]

That line of reasoning has merit, but it would not address another form of panic that is evident in the weather-discussion forums that have become increasingly popular. Yes, some people panic when it *doesn't* snow.

OCD and snOwCD

The thread appeared in the New England branch of the American Weather forum. The opening post read, "Winter cancel." The title of the thread: "Panic Room."

This was on Christmas Eve!

"Can't cancel what has only just begun," a forum moderator responded. At that point winter was less than 60 hours old.[12]

As it turned out, that first poster was onto something regarding the winter of 2019–2020, which speaks to one reason these chat boards draw so many eyes. Yes, outrageous and uninformed comments are inevitable. Yet, I have often found that the discussions are quite informed and the authors are way out ahead of the conventional forecasts and outlooks. I go there first before I check the National Weather Service and the commercial forecasts.

Some of the frequenters are working meteorologists employed in the private sector who offer insights, analyses, and opinions that they wouldn't express publicly. Some of the nonmeteorologists are at least as knowledgeable as the pros.

That would be the result of the incredible volume of data and information now in the public domain—computer-model runs from the United States, the European Community, the United Kingdom, Canada, and Japan; "oscillation" indexes of barometric-pressure differences from all over the planet; internal, detailed forecast discussions; and, above all, the pursuit of snow with an obsessed detective's passion. Frustration at not finding it would be a natural byproduct.

Such passion might well be unfathomable to the normal population, but it's real, and for true snow-lovers, missing out on a promised storm, especially one that creams some place not so far away, is sheer heartbreak, if not out-right grief. Intensifying the frustration is that with the chat boards and the published snow totals, it is possible to know who is and who is not getting what, how a "gravity wave" might be burying a town 10 miles away while robbing you of your snow.

As a threatened January storm was achingly slow to materialize in the Philadelphia region during a painful winter for the snow seekers in 2020, a poster on Phillywx.com read, "We watch and wait. I'll hoist a preemptive WSW (Weenie Suicide Watch), primarily for myself . . . winter is over, woe is me, bring on spring."[13]

It also was a nonevent in New York. Counseling a heartbroken snow-lover who had written, "I lost a lot of sleep this week tracking this," a poster responded, "I hope you can stop doing this, for your health's sake."[14]

Some panic when it snows; others panic when it *doesn't*!?

I often remind our younger reporters that they should never forget that they are covering a peculiar species. Finding logic in humans' relationship with snow might be akin to looking for a dime buried in an avalanche. Yet, I would argue that the two panics well from one source, as "two eyes that make one in sight," in the words of Robert Frost.[15]

I believe I understood this paradox intuitively on that February night, when I was nine years old experiencing an all-time favorite snowstorm.

What We Want

That February night when I was nine, I obeyed Bill Morgan's unstated com-mand and headed home. I was able to walk in the middle of sound-smothered Crosby Street, luxuriating in the privilege of being the first to step in snow touched only by the fresh, fast-falling flakes sculpted artfully and reshaped by the powerful wind. The vague fright borne on the stinging air and unprec-edented sense of solitude strangely intensified the exhilaration.

How different Crosby Street was on this night, and not only physically. The sun had no relationship with Crosby Street, which was our bleak play-ground, shared with the sparse traffic. Even on summer afternoons, it was blackened by shadows; the mammoth Chester Rescue Mission building on the west side of the street barricaded the light from the narrow houses on the east side. By day, we played football and baseball among the parked cars, the balls clunking off car hoods and roofs. As darkness settled on the street, the ball games yielded to hide 'n' seek and the taunting of the unfortunate men who had taken shelter at the mission. We would yell into the windows as they sang hymns after supper.

On this February evening, the games, taunts, and shadows were buried beneath the snow and hurled away by the wind; the mission men were invisible, and not just them.

It was not unusual for a nine-year-old to go outside alone in my neighborhood in those days. Unwitting adult chaperones roamed about, walking to and from the store or beer garden, their faces glowing with the cold or consumed alcohol. I was never alone—until this night. It was as though the snow had entombed everyone and everything inside the houses, and imposed an interdict on the cars and all life.

I entered the alley that connected Crosby Street with my backyard. I marveled at the radical purity of the alley, at the way snow concealed even the rat-hole that was large enough to trap an ankle, at the flakes coating my coat, at how the new-fallen snow had erased my boot prints, at how it silenced the absent dogs whose vicious barking would rattle the fences that lined the alley. It took me a solid half-hour to get to my backyard, less than a block from Crosby.

Under the spell of snow, the alley achieved a profound pastoral perfection, the spindrift swirling waves exalting atop luxurious layers of white. For one night, my relationship with this uninviting passageway was redefined. For one night, we could start over.

If I had to pick the unlikeliest venues for an encounter with perfection, it would be that sad, hideous, and relentlessly tedious corridor with the bark-rattled fence boards and cavernous rat-hole. That it could happen there is eloquent and persuasive testimony to snow's transformative powers. This is the memory that has survived the bulldozed houses and evaporated neighborhood. Snow.

As I walked along the narrow corridor from the backyard to the front porch, I admired the ingenuity of the snow, how it squeezed through the tight roof gaps to cover the narrow brick walkway between our house and the building next door. I was unaware that the ingenious snow had fallen upon and etched its way into the miraculous recesses of memory.

I entered our vestibule, red-cheeked, damp, and frozen, a thick and hard-frozen coating of snow covering my pant-legs. It must have been about eight o'clock. I shook off my coat and let the warmth of the house melt the snow on my pants.

By the time I entered the living room, about 10 feet from the vestibule door, the liquid remnants of the snow that had lodged inside my boots had saturated my socks. I left a trail of wet sock prints across the green tiles as I approached the prime snow-watching station of the house—an east-facing Victorian scale window adjacent to a radiator in the far end of the room. I don't recall who was watching an unremembered program on the flickering TV on the other side of the room. I pulled a chair close to the radiator and raised the Venetian blind on the window. I positioned the chair for the perfect sight line, in which a porch post blocked out the glare of the street lamp, giving me an unobstructed view of falling snow.

Outside, something had changed. On the east-facing window I could feel the icy wind that was agitating the thickly falling snow. The storm had crossed a barrier; it was no longer recreational snow. No cars dared test the thickening white barricade on our normally busy street. No one walked past the window. All was silent, save for the subtle whistling of the gusts that shouldered against the glass. I had no choice but to stay inside, to admire, in awe of the white fury, to watch a world forced to accept nature's terms. On the warm, rounded metal of the radiator I placed the soggy sock bottoms that covered my soaked soles. Blissfully I could have spent the rest of my life warming my feet on that radiator and hailing the conqueror in all its ferocity.

Revisiting that night, I know now that with the clarity of a child's intuition I had come to understand the number-one reason that snow is such an obsession for so many of us. It isn't fear of isolation. Quite the contrary. I believe that far more of us who obsess about snow do not fear isolation at all. That profound sense of temporary isolation, all the more pleasurable because we know it can't last long, is precisely what so many of us crave. The fact that we know that it can't last long makes it all the more pleasurable. It is an almost universal desire among snow-lovers. Even in places where snow is on the ground most of the winter, in Chicago and Minneapolis, they still want that Big One, the Grand Isolator. The craving is at its most powerful in those metropolitan centers where that desire is requited ever so rarely. The very remoteness of the threat serves only to intensify the yearning, elevating the prospect of snow isolation to the level of the great white hope. With an inextinguishable pilot light, it is a hope readily rekindled by a rumor of snow or a computer-model run that promises the storm of the century, or the sight of actual flakes.

More than panic, the supermarket rush is about hope: hope that we will have no choice but to sit inside and enjoy the harvest of our preparations. When the snow is deep and raging, and incited by blizzard winds, that sensation of well-being is unmatchable. We have no choice but to admire the snow and luxuriate in the limitations it imposes, blessedly unburdened of free will. For a magical period, in that "tumultuous privacy of storm,"[16] we live the fantasy of being Meta Stern at dawn on March 12, 1888, on the Lower East Side of New York.

This transcendent sensation is heightened to an immeasurable degree by the inescapable reality that what we are watching is, in the end, incandescent. Life endures, snow melts, but for an enchanting interlude we can believe that the snow has conquered all those forbidding alleys.

All the while, as we watch, at our deepest levels, we know that the "simplicity" of the snow is deceptive. That the "simplicity" we see in the falling whiteness is truly an illusion, artfully dispelled by an aftermath of ingenious and ubiquitous whiteness wind-shaped into forms that no human hand could replicate.

"Come see the north wind's masonry," Emerson writes of snowdrifts. "So fanciful, so savage."[17]

The flakes, themselves, are miracles of design, intricate beyond our comprehension, reminders of the beautifully mystical nature of our own beings and our place in nature, "God's DNA."

Addendum

The following are the 15 snowiest U.S. cities with a population of 100,000 or more for which National Weather Service official snow records (in inches) are available. The top three for normal seasonal snowfall are in the New York Snow Belt; Buffalo is only number three. Massachusetts has one city on the list, and it's not Boston. (Denver's totals are based on the 30-year period ending with the winter of 1994–1995.)

Syracuse, New York	123.8
Rochester, New York	99.5
Buffalo, New York	94.7
Grand Rapids, Michigan	74.9
Anchorage, Alaska	74.5
South Bend, Indiana	66.6
Worcester, Massachusetts	64.1
Denver, Colorado	60.3
Cleveland, Ohio	58.1
Salt Lake City, Utah	56.2
Billings City, Montana	55.0
Minneapolis, Minnesota	54.4
Green Bay, Wisconsin	51.4
Madison, Wisconsin	50.9

Notes

Chapter One

1. Christopher Columbus, *Journal of the First Voyage of Columbus, American Journeys Collection, Wisconsin Historical Society Digital Library and Archives*, 2003, p. 142, http://www.americanjourneys.org/pdf/AJ-062.pdf (accessed March 12, 2019).

2. Columbus, *Journal of the First Voyage of Columbus*, p. 119.

3. David Ludlum, *Early American Winters, 1604–1820* (Boston: American Meteorological Society, 1966), p. 6.

4. Ludlum, *Early American Winters*, p. 243.

5. "The Ice Man, a Cherokee Legend," *First People, American Indian Legends*, https://www.firstpeople.us/FP-Html-Legends/TheIceMan-Cherokee.html (accessed March 12, 2019).

6. Christopher Columbus, *The First Letter of Christopher Columbus to the Nobel Lord Raphael Sanchez* (Boston: Trustees of the Boston Public Library, 1891), p. 7.

7. Ludlum, *Early American Winters*, p. 7.

8. William Wood, *New England Prospects, Internet Archive*, pp. 3–4, https://archive.org/details/woodsnewengland00woodgoog/page/n53 (accessed March 12, 2019).

9. Ludlum, *Early American Winters*, pp. 17–18.

10. Ludlum, *Early American Winters*, p. 19.

11. John Winthrop, *History of New England, 1630–1649*, ed. James K. Hosmer (New York: Charles Scribner's Sons, 1908), p. 269.

12. David Ludlum, *The Weather Factor* (Boston: American Meteorological Society, 1984), p. 15.

13. Winthrop, *History of New England*, p. 90.

14. Thomas Hutchinson, *The History of the Province of Massachusetts Bay Province from the Charter of King William and Mary in 1691, Until the Year 1750*, 2 vols. (London: J. Smith, 1758), p. 101.

15. Cotton Mather, *Diary of Cotton Mather, 1681–1724* (Boston: Massachusetts Historical Society, 1911), p. 212.

16. Ludlum, *Early American Winters*, p. 16.

17. Ludlum, *Early American Winters*, p. 17.

18. Ludlum, *Early American Winters*, p. 41.

19. Ludlum, *Early American Winters*, p. 42.

20. Cotton Mather, *Selections from Cotton Mather*, Kenneth B. Murdock (New York: Harcourt Brace, 1926), p. 374.

21. Mather, *Selections from Cotton Mather*, pp. 374–75.

22. Mather, *Diary of Cotton Mather*, p. 440.

23. Ludlum, *Early American Winters*, p. 245.

24. Ludlum, *Early American Winters*.

25. Ludlum, *Early American Winters*, pp. 144–46.

26. Ludlum, *Early American Winters*, p. 147.

Chapter Two

1. Hannah Fry and Alejandro Reyes-Velarde, "Snow Comes to L.A., with Powder in Malibu, Pasadena, West Hollywood," *Los Angeles Times*, February 21, 2019, https://www.latimes.com/local/lanow/la-me-california-snow-20190221-story.html (accessed March 9, 2019).

2. "Freak Southland Storm Brings Snow, Sleet, Hail," *Los Angeles Times*, January 22, 1962, p. 1.

3. Kenneth G. Libbrecht, "What's It Good For," *Snowcrystals.com*, http://www.snowcrystals.com/motivation/motivation.html (accessed March 9, 2019).

4. Kenneth G. Libbrecht, interview with author, February 5, 2019.

5. Libbrecht, interview with author.

6. "Lens Crafters," *American Physical Society News* 13, no. 3 (March 2004): 2, https://www.aps.org/publications/apsnews/200403/upload/mar04.pdf (accessed March 4, 2020).

7. John J. Fahie, *Galileo: His Life and Work* (New York: James Pott & Company, 1903), p. 208.

8. Charles Singer, "Notes on the Early History of Microscopy," *Proceedings of the Royal Society of Medicine* 7 (May 1, 1914): 261–62, https://journals.sagepub.com/doi/pdf/10.1177/003591571400701617 (accessed March 9, 2019).

9. Johannes Kepler, *The Six-Cornered Snowflake*, trans. Jacques Bromberg (Philadelphia, PA: Paul Dry Books, 2010), location 520, Kindle.

10. Kenneth G. Libbrecht, "Snowflake History," *Snowcrystals.com*, http://www.snowcrystals.com/history/history.html (accessed March 9, 2019).

11. Francis Charles Frank, "Early Discoverers XXXI: Descartes's Observations on the Amsterdam Snowfalls of 4, 5, 6, and 9 February 1635," *Journal of Glaciology* 13, no. 3 (1974): 535–39, https://doi.org/10.3189/S0022143000023261.

12. Samuel Pepys, *Diary of Samuel Pepys*, vol. 2., ed. Henry B. Wheatley (New York: Macmillan and Company, 1894), p. 316.

13. Robert Hooke, *Micrographia: Or Some Physiological Descriptions of Minute Bodies Made by Magnifying Glass* (Lincolnwood, IL: Science Heritage Ltd., 1987), preface, https://play.google.com/books/reader?id=LsbBada4VVYC&hl=en&pg=GBS .PP7 (accessed July 11, 2020).

14. Hooke, *Micrographia*, Observation 59, p. 204.

15. Hooke, *Micrographia*, Observation 14, p. 91.

16. Stephen Schneider and Randi Londer, *The Coevolution of Climate and Life* (San Francisco, CA: Sierra Club Books, 1984), p. 114.

17. Hooke, *Micrographia*, Observation 14, p. 91.

18. Sue Richardson, phone interview with author, December 14, 2019.

19. Duncan C. Blanchard, *The Snowflake Man: A Biography of Wilson A. Bentley* (Blacksburg, VA: McDonald & Woodward), p. 22.

20. Blanchard, *The Snowflake Man*, p. 22.

21. Blanchard, *The Snowflake Man*, p. 116.

22. Gloria May Stoddard, *Snowflake Bentley: Man of Science, Man of God* (Shelburne, VT: New England Press, 1985).

23. Wilson A. Bentley, "Photographing Snowflakes," *Popular Mechanics Magazine*, vol. 37, 1922, pp. 309–12.

24. Stoddard, *Snowflake Bentley*.

25. Richardson, phone interview with author.

26. Blanchard, *The Snowflake Man*, p. 27.

27. Stoddard, *Snowflake Bentley*.

28. Mary B. Mullet, "The Snowflake Man," from *American Magazine*, republished by the Jericho Historical Society, https://snowflakebentley.com/mary-mullet-article (accessed July 11, 2020).

29. Richardson, phone interview with author.

30. Blanchard, *The Snowflake Man*, pp. 30–32.

31. Stoddard, *Snowflake Bentley*.

32. Mullet, "The Snowflake Man."

33. Richardson, phone interview with author.

34. James H. Powers, "Fame Comes to Snowflake Bentley after 35 Years," *Sunday Boston Globe*, January 2, 1921, p. 54.

35. J. E. Wolff, "Exhibition and Preliminary Account of a Collection of Microphotographs of Snow Crystals Made by Wilson A. Bentley," *Proceedings of the American Academy of Arts and Sciences* 23, no. 23 (1898): 431–32.

36. Stoddard, *Snowflake Bentley*.

37. Richardson, phone interview with author.

38. Wilson A. Bentley and George H. Perkins, "A Study of Snow Crystals," *Appleton's Popular Science Monthly* 53 (1898): 75.

39. Bentley and Perkins, "A Study of Snow Crystals," pp. 75–82.

40. Stoddart, *Snowflake Bentley*, p. 109.

41. Stoddart, *Snowflake Bentley*, p. 109.

42. Margaret Wertheim and Kenneth Libbrecht, "Building a Better Snowflake: An Interview with Kenneth Libbrecht," *Cabinet* 29 (Spring 2008), http://www.cabinetmagazine.org/issues/29/wertheim.php (accessed March 9, 2019).

43. John F. Fuller, *Thor's Legions: Weather Support to the U.S. Air Force and Army, 1937–1987* (Boston: American Meteorological Society, 1990).

44. Kenneth G. Libbrecht and Rachel Wing, *The Snowflake: Winter's Frozen Artistry* (Minneapolis, MN: Voyageur Press, 2015).

45. Wertheim and Libbrecht, "Building a Better Snowflake."

46. Herbert B. Nichols, "Book Reviews: *Snow Crystals: Natural and Artificial*," by Ukichiro Nakaya. *Science* 120, no. 3,123, November 5, 1954, p. 755, http://science.sciencemag.org/content/120/3123/755.1 (accessed March 9, 2019).

47. Libbrecht, interview with author.

48. "The First Artificial Snowflake," *Asia Research News*, June 13, 2018, https://www.asiaresearchnews.com/html/article.php/aid/11800/cid/2/research/science/hokkaido_university/the_first_artificial_snowflake_%5Basia_research_news_2018_moments_in_history%5D.html (accessed July 17, 2020).

49. Henry David Thoreau, *The Journal, 1837–1861*, ed. Damion Searls (New York: New York Review Books, 2009), p. 356.

50. Libbrecht, interview with author.

51. Libbrecht, interview with author.

52. Libbrecht, interview with author.

53. Paul J. Kocin and Louis Uccellini, *Northeast Snowstorms: The Cases*, vol. 2 (Boston: American Meteorological Society, 2004): 193–94.

54. Kenneth G. Libbrecht, curriculum vitae.

55. Wertheim and Libbrecht, "Building a Better Snowflake."

56. Wertheim and Libbrecht, "Building a Better Snowflake."

57. Wertheim and Libbrecht, "Building a Better Snowflake."

58. Libbrecht, interview with author.

59. Phillipe M. Binder, "Book Reviews: *Snowflakes: Winter's Secret Beauty*," Kenneth G. Libbrecht. *American Journal of Physics* 72, no. 1,134 (August 2004), https://doi.org/10.1119/1.1764565.

60. Libbrecht, interview with author.

61. Kenneth G. Libbrecht, "Is It Really True That No Two Snowflakes Are Alike?" *Snowcrystals.com*, https://www.its.caltech.edu/~atomic/snowcrystals/alike/alike.htm (accessed March 9, 2019).

62. Motoi Kumai, "Electron Microscope Study of Snow-Crystal Nuclei," *J. Meteor* 8, no. 3 (1951): 151–56, https://doi.org/10.1175/1520-0469(1951)008<0151:EMSOSC>2.0.CO;2.

63. Wilson A. Bentley, "The Snowflake Man," *Snowflakebentley.com*, http://snow flakebentley.com/bio.htm (accessed March 9, 2019).

64. Mullet, "The Snowflake Man."

65. Richardson, phone interview with author.

66. Richardson, phone interview with author.

67. Blanchard, *The Snowflake Man*, p. 34.

Chapter Three

1. Blake McKelvey, *Snow in the Cities: A History of the Urban Response* (Rochester, NY: University of Rochester Press, 1995), p. 22.

2. Hector St. John de Crevecoeur, *Letters from an American Farmer*, edited by Dennis D. Moore (Athens: University of Georgia Press, 1995), p. 148.

3. McKelvey, *Snow in the Cities*, pp. 20–23.

4. Francis F. Brown, ed., *Golden Treasury of Poetry and Prose* (New York and St. Louis, MO: N. D. Thompson and Company, 1883), p. 852.

5. Carmen Nigro, "So, Why Do We Call It Gotham, Anyway?" *New York Public Library*, January 25, 2011, https://www.nypl.org/blog/2011/01/25/so-why-do-we-call -it-gotham-anyway (accessed March 9, 2019).

6. Artis Q. Wright, "Designing the City of New York: The Commissioners' Plan of 1811," *New York Public Library*, July 30, 2010, https://www.nypl .org/blog/2010/07/30/designing-city-new-york-commissioners-plan-1811 (accessed March 9, 2019).

7. Robin Nagle, *Picking Up: On the Streets and Behind the Trucks with the Sanitation Workers of New York City* (New York: Farrar, Straus and Giroux, 2013).

8. John F. Stover, *History of the Baltimore and Ohio Railroad* (West Lafayette, IN: Purdue University Press, 1987).

9. Stover, *History of the Baltimore and Ohio Railroad*, p. 35.

10. Jon Nese and Glenn Schwartz, *The Philadelphia Area Weather Book* (Philadelphia, PA: Temple University Press, 2002).

11. Nese and Schwartz, *The Philadelphia Area Weather Book*.

12. Paul J. Kocin et al., "Overview of the 12–14 March 1993 Superstorm," *Bulletin of the American Meteorology Society* 76 (February 1995): 166.

13. Benjamin Silliman, *The American Journal of Arts and Sciences*, vol. 20 (New Haven, CT: Hezekiah Howe, 1831).

14. "Local Intelligence, City Nuisances," *New York Times*, December 18, 1865, p. 2.

15. Nagle, *Picking Up*.

16. McKelvey, *Snow in the Cities*, p. 26.

17. *The Commissioners of Patents Journal*, no. 367, July 10, 1857, https://play .google.com/books/reader?id=fZJGAQAAMAAJ&hl=en&pg=GBS.PA881 (accessed March 9, 2019).

18. McKelvey, *Snow in the Cities*.

19. "Another Great Snowstorm," *New York Daily Times*, January 20, 1857, p. 8.

20. "Another Great Snow-Storm, Spring Nipped in the Bud," *New York Times*, March 3, 1857, p. 8.

21. "Salting the Streets," *New York Times*, January 31, 1860.

22. "The Snow-Storm," *New York Times*, January 5, 1859, p. 1.

23. McKelvey, *Snow in the Cities*, pp. 33–34.

24. McKelvey, *Snow in the Cities*.

25. "Rotary Snowplow as Used on a German Railroad," *Railway Review* 66 (1920): 69.

26. "Melting Away the Snow," *New York Times*, January 7, 1881, p. 2.

27. "Board of Health," *New York Times*, February 2, 1871, p. 6.

28. McKelvey, *Snow in the Cities*.

29. *Decisions of the Courts of New York*, vol. 1, with notes by Austin Abbott (New York: Ward and Peloubet, 1877), pp. 75–80.

30. Bogdan Horbal, "The Early Proposed Railways for New York City, Part 2," *New York Public Library*, July 7, 2015, https://www.nypl.org/blog/2015/07/07/early-proposed-railways-nyc-2 (accessed March 9, 2019).

31. Nagle, *Picking Up*.

Chapter Four

1. Judd Caplovich, *Blizzard! The Great Storm of '88* (Vernon, CT: VeRo, 1987), p. 3.

2. Meta Stern Lillenthal, *Dear Remembered World: Childhood Memories of an Old New Yorker* (New York: Richard R. Smith, 1947), pp. 221–30.

3. David Laskin, *The Children's Blizzard* (New York: Harper Perennial, 2004).

4. H. L. Mencken, *The American Language* (New York: Alfred A. Knopf, 2006), p. 219.

5. David Ludlum, *Early American Winters, 1604–1820* (Boston: American Meteorological Society, 1966), p. 240.

6. "Blizzard," *National Weather Service*, https://w1.weather.gov/glossary/index.php?letter=b (accessed March 9, 2019).

7. "The Storm and City Transit," *New York Times*, March 13, 1888, p. 4.

8. "In a Blizzard's Grasp," *New York Times*, March 13, 1888, p. 1.

9. Mary Cable, *The Blizzard of '88* (New York: Atheneum, 1988), pp. 32–33.

10. Paul J. Kocin, "An Analysis of the Blizzard of 1888," *Bulletin of the American Meteorological Society* 64, no. 11 (November 1, 1983): 1,258–72.

11. "On this Day," *New York Times*, March 3, 1888, http://movies2.nytimes.com/learning/general/onthisday/harp/0303.html (accessed March 9, 2019).

12. Blake McKelvey, *Snow in the Cities: A History of the Urban Response* (Rochester, NY: University of Rochester Press, 1995), p. 59.

13. Cable, *The Blizzard of '88*, p. 99.

14. Caplovich, *Blizzard!*

15. McKelvey, *Snow in the Cities.*

16. McKelvey, *Snow in the Cities.*

17. "Towards a Cleaner City," *Virtual New York, City University of New York,* https://virtualny.ashp.cuny.edu/blizzard/sanitation/san2.html (accessed March 9, 2019).

18. John L. Sprague and Joseph J. Cunningham, "A Frank Sprague Triumph: The Electrification of Grand Central Terminal," *IEEE Power and Energy Magazine,* January–February 2013, http://magazine.ieee-pes.org/january-february-2013/history-6/ (accessed March 9, 2019).

19. Sprague and Cunningham, "A Frank Sprague Triumph."

20. Sprague and Cunningham, "A Frank Sprague Triumph."

21. H. S. Stidham, *The Problem of Snow Removal* (New York: Martin B. Brown Company, 1897), p. 4.

22. Stidham, *The Problem of Snow Removal.*

23. Stidham, *The Problem of Snow Removal,* pp. 11–13.

24. Bob DiFlorio, interview with author, *Philadelphia Inquirer,* February 10, 1994.

25. McKelvey, *Snow in the Cities.*

26. "The Ruggles Snow Plow," *Electrical Age,* January 2, 1897, p. 14.

27. Fred House, phone interview with author, March 2000.

28. McKelvey, *Snow in the Cities.*

29. McKelvey, *Snow in the Cities.*

30. Phil Wright, interview with author, *Philadelphia Inquirer,* February 10, 1994.

31. "The Salt Nuisance on Railroad Tracks," *New York Times,* February 1, 1873, p. 8.

32. "Snow Removal," *Journal of the American Society of Mechanical Engineers* 37 (January–December 1915): 92–95.

33. McKelvey, *Snow in the Cities.*

34. John C. Wood and Michael C. Wood, eds., *Henry Ford: Critical Evaluations in Business and Management,* vol. 1 (London and New York: Routledge/Taylor and Francis, 2002), p. 2.

35. "State Motor Vehicle Registrations, by Years, 1900–1995," *Federal Highway Administration,* https://www.fhwa.dot.gov/ohim/summary95/mv200.pdf (accessed March 9, 2019).

36. Bernard Mergen, *Snow in America* (Washington, DC: Smithsonian Institution Press, 1997).

37. McKelvey, *Snow in the Cities.*

38. National Research Council, Division on Earth and Life Sciences, Transportation Research Board, Board on Environmental Studies and Toxicology, and Committee on Ecological Impacts of Road Density, *Assessing and Managing the Ecological Impacts of Paved Roads* (Washington, DC: National Academies Press, 2005).

39. Mergen, *Snow in America.*

40. Mergen, *Snow in America,* p. 61.

41. Anthony R. Wood, "The Salt Is Ready," *Philadelphia Inquirer Magazine*, December 25, 1994.

42. "Salt Production in Syracuse, NY, the Salt City, and the Hydrogeology of the Onondaga Creek Valley," *Department of the Interior, U.S. Geological Survey*, November 2000, https://pubs.usgs.gov/fs/2000/0139/report.pdf (accessed March 9, 2019).

43. Arthur H. Grant and Harold S. Buttenheim, eds., *The American City*, vol. 25 (New York: Buttenheim, 1921).

44. Frank Kummer, "Road Salt Levels in Some Philadelphia-Area Streams Hit Toxic Levels," *Inquirer.com*, May 28, 2019, https://www.inquirer.com/news/delaware-river-watershed-salt-levels-20190528.html (accessed January 4, 2020).

45. Molly Hunt et al., "Chlorides in Freshwater," *University of Rhode Island, College of Environment and Life Sciences*, March 2012, http://cels.uri.edu/docslink/ww/water-quality-factsheets/Chlorides.pdf (accessed January 4, 2020).

46. Samantha Briggs, phone interview with author, January 3, 2020.

47. Briggs, phone interview with author.

48. "Road Salt and Water Quality," *New Hampshire Department of Environmental Services*, 2016, https://www.des.nh.gov/organization/commissioner/pip/factsheets/wmb/documents/wmb-4.pdf (accessed March 9, 2019).

49. Xianming Shi, phone interview with author, November 2019.

50. Briggs, phone interview with author.

51. "Pass the Salt," *ForConstructionPros.com*, January 12, 2015, https://www.forconstructionpros.com/asphalt/article/12022723/the-only-time-vodka-roads-should-mix (accessed January 4, 2020).

52. Anthony R. Wood, "To Treat Icy Roads, Highway Agencies Look to Grapes, Cheese, and Vodka as Alternatives to Salt," *Inquirer.com*, December 16, 2019, https://www.inquirer.com/weather/snow-ice-roads-winter-driving-salt-alternatives-beets-grapes-vodka-20191216.html (accessed January 15, 2020).

53. Medhi Honarvar Nazari and Xianming Shi, "Developing Renewable Agro-Based Anti-Icers for Sustainable Winter Road Maintenance Operations," *Journal of Materials in Civil Engineering* 31, no. 2 (December 2019), https://ascelibrary.org/doi/10.1061/%28ASCE%29MT.1943-5533.0002963 (accessed January 15, 2020).

54. Wood, "To Treat Icy Roads."

55. Ann Fordock, phone interview with author, March 2019.

56. Wood, "To Treat Icy Roads."

57. McKelvey, *Snow in the Cities*.

58. Wallace Bolen, "Salt," *U.S. Geological Survey 2015 Minerals Yearbook* (Reston, VA: U.S. Geological Survey, 2015).

Chapter Five

1. Cora Conner, phone interview with author, May 2010.

2. Leda Hunter, personal remembrance of a Bly, Oregon, resident, date unknown.

3. Bert Webber, *Silent Siege* (Fairfield, WA: Ye Galleon Press, 1984), p. 227.

4. William M. Tuttle Jr., *Daddy's Gone to War: The Second World War in the Lives of America's Children* (New York and Oxford, UK: Oxford University Press, 1993), p. 10.

5. Jack Smith, phone interview with author, May 2010.

6. "General Report No. 3 on Free Balloons and Related Incidents" (declassified), Record Group No. 77, Military Intelligence Division, War Department, National Archives, April 3, 1945.

7. Robert C. Mikesh, *World War II Balloon Bomb Attacks upon North America* (Washington, DC: Smithsonian Institution Press, 1973).

8. Gerald W. Williams, "Balloon Bombs of World War II," U.S. Forest Service report, April 13, 2004.

9. Mikesh, *World War II Balloon Bomb Attacks upon North America.*

10. Mikesh, *World War II Balloon Bomb Attacks upon North America.*

11. Mikesh, *World War II Balloon Bomb Attacks upon North America,* 28–29.

12. Mikesh, *World War II Balloon Bomb Attacks upon North America.*

13. John M. Lewis, "Ooishi's Observations Viewed in the Context of Jet Stream Discovery," *Bulletin of the American Meteorological Society* (March 2003): 357–65, https://journals.ametsoc.org/doi/10.1175/BAMS-84-3-357 (accessed March 9, 2019).

14. Yoshitomo Kojoh, personal correspondence with author, translated by Akira Suwa, May 2010.

15. Rebecca Maskel, "Why Was the Discovery of the Jet Stream Mostly Ignored?" *Air and Space Magazine,* April 2018, https://www.airspacemag.com/as-next/as-next-may-unbelievablebuttrue-180968355/ (accessed March 16, 2019).

16. Kojoh, personal correspondence with author.

17. Reid A. Bryson, "The Discovery of the Jet Stream," *Wisconsin Academy Review* (Summer 1994): 16. Manuscript courtesy of the University of Wisconsin.

18. Norman A. Phillips, "Carl Gustaf Rossby, His Times, Personality, and Actions," *Bulletin of the American Meteorological Society* (June 1998): 1,101, https://journals.ametsoc.org/doi/pdf/10.1175/1520-0477%281998%29079%3C1097%3ACGRHTP%3E2.0.CO%3B2 (accessed March 16, 2019).

19. John M. Lewis, Matthew G. Fearon, and Harold Klieforth, "Herbert Riehl, Intrepid and Enigmatic Scholar," *Bulletin of the American Meteorological Society* (July 2012): 963–85, https://journals.ametsoc.org/doi/pdf/10.1175/BAMS-D-11-00224.1 (accessed March 16, 2019).

20. Andrew C. Winters, "The Jet Stream: The Atmosphere's Interstate Highway," *State University of New York at Albany, Department of Atmospheric and Environmental Sciences,* November 10, 2016, http://www.atmos.albany.edu/facstaff/awinters/JetPrimer.pdf (accessed March 16, 2019).

21. Lewis, Fearon, and Klieforth, "Herbert Riehl," 971.

22. Richard A. Keen, *Skywatch East: A Weather Guide* (Golden, CO: Fulcrum, 1989).

23. Herbert Riehl, "Jet Streams of the Atmosphere," *Department of Atmospheric Science, Colorado State University,* Report No. 32 (May 1962): 1–2,

https://mountainscholar.org/bitstream/handle/10217/24425/0032_Bluebook
.pdf?sequence=1&isAllowed=y, (accessed March 16, 2019).

24. Robert Kunzig, *The Restless Sea: Exploring the World Beneath the Waves* (New York and London: W. W. Norton and Company, 1999), p. 287.

25. Thomas F. McIlwraith and Edward K. Muller, eds., *North America: The Historical Geography of a Changing Continent* (Lanham, MD: Rowman & Littlefield, 2001), p. 38.

26. Henry Stommel, *The Gulf Stream: A Physical and Dynamical Description* (Berkeley: University of California Press, 1960).

27. Benjamin Franklin, *The Ingenious Dr. Franklin: Selected Scientific Letters of Benjamin Franklin*, ed. Nathan Goodman (Philadelphia: University of Pennsylvania Press, 1974), p. 131.

28. Bill Johns, interview with author, October, 2005.

29. "Tidal Power: Florida's Ocean Current Potential," *Future Power Technology Magazine* November 12, 2014, https://www.power-technology.com/features/featuretidal-power-floridas-ocean-current-potential-4379164/ (accessed March 2019).

30. Paul Kocin, interview with author, *Philadelphia Inquirer*, December 19, 2005.

31. Samuel Eliot Morison, "Christopher Columbus, Mariner," *American Heritage* 7, no. 1 (December 1955), https://www.americanheritage.com/christopher-columbus-mariner (accessed March 16, 2019).

32. Christopher Columbus, "Excerpts from the Christopher Columbus Log, 1492 A.D.," *Scholasticum*, https://www.franciscan-archive.org/columbus/opera/excerpts.html (accessed March 16, 2019).

33. Henry Stommel, "The Westward Intensification of Wind-Driven Ocean Currents," *Eos* 29, no. 2 (1948): 202–6, https://agupubs.onlinelibrary.wiley.com/doi/abs/10.1029/TR029i002p00202 (accessed March 9, 2019).

34. Frederick Sanders and John Gyakum, "Synoptic-Dynamic Climatology of the 'Bomb,'" *Monthly Weather Review* (October 1980): 1,589.

35. Nicole Mortillaro, "Meet the Canadian Who Helped Coin the Term 'Weather Bomb,'" *CBC News*, January 5, 2018, https://www.cbc.ca/news/technology/canadian-coined-term-weather-bomb-1.4474431 (accessed March 15, 2019).

36. Fred Otsby, during talk at a Washington media conference, December 1983.

37. Frederick P. Shuman, "History of Numerical Weather Prediction at the National Meteorological Center," *Bulletin of the American Meteorological Society* (April 1989): 286.

38. Norman A. Phillips, *Biographical Memoirs*, vol. 66 (Washington, DC: National Academies Press, 1995), https://www.nap.edu/read/4961/chapter/6p (accessed March 16, 2019).

39. Norman A. Phillips, "Jule Charney's Influence on Meteorology," *Bulletin of the American Meteorological Society* (May 1982): 492–98, https://journals.ametsoc.org/doi/pdf/10.1175/1520-0477%281982%29063%3C0492%3AJCIOM%3E2.0.CO%3B2 (accessed March 16, 2019).

40. J. R. Fulks, "The Early November Snowstorm of 1953," *Weatherwise* 7, no. 1 (1954): 12–16, doi:10.1080/00431672.1954.9930307.

41. Jay S. Winston, "The Weather and Circulation of November 1953," *Monthly Weather Review* (December 1953): 370.

Chapter Six

1. Anthony R. Wood, "How Weather Forecasters Got Snowed," *Philadelphia Inquirer*, January 26, 2000, p. A12.

2. Gary Szatkowski, interview with author, *Philadelphia Inquirer*, January 25, 2000.

3. Elliot Abrams, interview with author, *Philadelphia Inquirer*, January 26, 2000.

4. Anthony R. Wood, "Raining Money," *Philadelphia Inquirer Magazine*, February 11, 2001, p. 10.

5. Greg Postel, interview with author, January 16, 2020.

6. Tom Topousis, "March Lion Turns into N.Y. Pussycat—but Talk of Big, Bad Blizzard Leaves Local Air Travel in Chaos," *New York Post*, March 6, 2001, https://nypost.com/2001/03/06/march-lion-turns-into-n-y-pussycat-but-talk-of-big-bad-blizzard-leaves-local-air-travel-in-chaos/ (accessed March 18, 2019).

7. David O'Reilly and Anthony R Wood, "Massive Snowstorm Heads toward Region; Up to 2 Feet Predicted," *Philadelphia Inquirer*, March 4, 2001, p. 1.

8. Monica Yant Kinney, "Storm Holds Off—for a Day," *Philadelphia Inquirer*, March 5, 2001, p. 1.

9. Sandy Hingston, "17 Things You Might Not Know (or Were Trying to Forget) about John Bolaris," *Philadelphia Magazine*, May 19, 2016, https://www.phillymag.com/news/2016/05/19/john-bolaris-american-greed/ (accessed July 15, 2020).

10. Victor Fiorillo, "One of Us: Glenn 'Hurricane' Schwartz," *Philadelphia Magazine*, December 16, 2011, https://www.phillymag.com/news/2011/12/16/philadelphia-weather-man-glenn-hurricane-schwartz-is-the-anti-john-bolaris/ (accessed January 21, 2019).

11. Glenn Schwartz, interview with author, *Philadelphia Inquirer*, March 6, 2001.

12. "Weather Forecasting, Banned by Taliban, Makes a Comeback in Afghanistan," *Canadian Press*, February 19, 2004.

13. "Edward Lorenz, Father of Chaos Theory and Butterfly Effect, Dies at 90," *MIT News*, April 16, 2008, http://news.mit.edu/2008/obit-lorenz-0416 (accessed March 18, 2019).

14. Andy Gregorio, interview with author, February 12, 1983.

15. Paul Kocin, phone interview with author, August 2005.

16. Chet Henricksen, interview with author, February 17, 2019.

17. "Superstorm of March 1993," *Natural Disaster Survey Report, National Oceanic and Atmospheric Administration*, May 1994, chapter 3, p. 4, https://www.weather.gov/media/publications/assessments/Superstorm_March-93.pdf (accessed March 18, 2019).

18. "The 1993 Storm of the Century," *National Weather Service*, https://www .weather.gov/tbw/93storm (accessed March 9, 2019).

19. "On This Day: The 1993 Storm of the Century," *National Centers for Environmental Information*, March 13, 2017, https://www.ncei.noaa.gov/news/1993-snow -storm-of-the-century (accessed March 9, 2019).

20. "Service Assessment, Blizzard of '96," *National Oceanic and Atmospheric Administration* (December 1996): 9, https://www.weather.gov/media/publications/assess ments/bz-mrg.pdf (accessed March 9, 2019).

21. Marc Santora and Emma G. Fitzsimmons, "New York City Is Spared Worst Effects of Snowstorm," *New York Times*, January 26, 2015, https://www.nytimes .com/2015/01/27/nyregion/new-york-blizzard.html (accessed March 9, 2019).

22. Angelo Cataldi, "Morning Show," 94WIP, January 26, 2015.

23. Louis W. Uccellini at a National Weather Service media teleconference, January 26, 2015.

24. Anthony R. Wood, "Figuring Out the Fizzled Forecast," *Philadelphia Inquirer*, January 28, 2015, p. A8.

25. Greg Carbin, interview with author, March 14, 2019.

26. Joe Miketta, interview with author, March 13, 2019.

27. Ross Dickman, e-mail to author, March 14, 2017.

28. Wood, "Figuring Out the Fizzled Forecast," p. A8.

29. Santora and Fitzsimmons, "New York City Is Spared Worst Effects of Snowstorm."

30. Chet Henricksen, interview with author, January 2017.

Chapter Seven

1. Joel N. Myers, interview with author, *Inquirer Magazine*, February 11, 2001.

2. Elliot Abrams, interview with author, April 1993.

3. Francis Davis, interview with author, June 1997.

4. "AccuWeather History," *Accuweather.com*, https://corporate.accuweather .com/history (accessed March 18, 2019).

5. Glenn Schwartz, interview with author, January 5, 2020.

6. "National Weather Service Enterprise Analysis Report," *National Weather Service*, June 8, 2017, p. 9, https://www.weather.gov/media/about/Final_NWS%20 Enterprise%20Analysis%20Report_June%202017.pdf (accessed March 18, 2019).

7. Joel N. Myers, phone interview with author, February 7, 1978.

8. Chet Henricksen, phone interview with author, February. 17, 2019.

9. Frank Batten, "Out of the Blue and into the Black," *Harvard Business Review*, April 2012, https://hbr.org/2002/04/out-of-the-blue-and-into-the-black (accessed March 18, 2019).

10. Frank Batten, with Jeffrey L. Cruikshank, *The Weather Channel: The Improbable Rise of a Media Phenomenon* (Cambridge, MA: Harvard Business School Press, 2002).

11. Amy Blitz, "Interview with Frank Batten," *Harvard Business School*, April 2000, http://www.hbs.edu/xentrepreneurs/pdf/frankbatten.pdf (accessed March 18, 2019).

12. Batten, *The Weather Channel*, p. 3.

13. Blitz, "Interview with Frank Batten."

14. Blitz, "Interview with Frank Batten."

15. Joel N. Myers, interview with author, June 2014.

16. "The Top 500 Sites on the Web," *Alexa*, https://www.alexa.com/topsites/category/News/Weather (accessed March 19, 2019).

17. "National Weather Service Enterprise Analysis Report: Findings on Changes in the Private Weather Industry," *Weather.gov*, June 8, 2017, p. 23, https://www.weather.gov/media/about/Final_NWS%20Enterprise%20Analysis%20Report_June%202017.pdf (accessed March 18, 2019).

18. Jason Samenow, "Weather Channel to Name Winter Storms: A Publicity and Power Play with Possible Value," *Washington Post*, October 2, 2012, https://www.washingtonpost.com/blogs/capital-weather-gang/post/weather-channel-to-name-winter-storms-a-publicity-and-power-play-with-possible-value/2012/10/02/efa49318-0c98-11e2-bb5e-492c0d30bff6_blog.html?utm_term=.b801a9c8b681 (accessed March 18, 2019).

19. Glenn Schwartz, phone interview with author, October 2012.

20. "Weather Channel Decision to Name Storms Will Cause Confusion," *AccuWeather*, October 3, 2012, https://www.accuweather.com/en/press/84520 (accessed March 18, 2019).

21. Tom Niziol, "The Science behind Naming Winter Storms at The Weather Channel," *weather.com*, October 3, 2014, https://weather.com/news/news/science-behind-naming-winter-storms-weather-channel-20140121 (accessed March 18, 2019).

22. Mary M. Glackin, "Naming Winter Storms: Time for the Weather Community to Cooperate," *Front Page, American Meteorological Society*, September 21, 2015, http://blog.ametsoc.org/columnists/naming-winter-storms-time-for-community-cooperation/ (accessed March 18, 2019).

23. Glackin, "Naming Winter Storms."

24. Samenow, "Weather Channel to Name Winter Storms."

25. Adam M. Rainear, Kenneth A. Lachlan, and Carolyn A. Lin, "What's in a #Name? An Experimental Study Examining Perceived Credibility and Impact of Winter Storm Names," *Weather, Climate, and Society* 9, no. 4 (October 2017): 815–22, https://journals.ametsoc.org/doi/10.1175/WCAS-D-16-0037.1 (accessed March 18, 2019).

26. AccuWeather news release, March 2, 2010, no longer posted.

27. Devin Leonard and Brian K. Sullivan, "Trump's Pick to Lead Weather Agency Spent 30 Years Fighting It," *Bloomberg Businessweek*, June 14, 2018, https://www.bloomberg.com/news/features/2018-06-14/trump-s-pick-to-lead-weather-agency-spent-30-years-fighting-it (accessed March 18, 2019).

28. Vince Stricherz, "Plan to Privatize Most Forecasting Would Cripple Weather Service," *University of Washington*, May 16, 2005, http://www.washington.edu/news/2005/05/16/plan-to-privatize-most-forecasting-would-cripple-weather-service-expert-says/ (accessed March 19, 2019).

29. Richard Hirn, "Nomination for Undersecretary of Commerce for Oceans and Atmosphere (NOAA Administrator)," *National Weather Service Employees Organization*, October 12, 2017, http://www.nwseo.org/Four%20Winds%202017/17_10_12_NWSEO_Myers_nomination_ltr.pdf (accessed March 18, 2019).

30. John Roach, "Can You Trust Weather Forecasts during the Government Shutdown? Absolutely!" *AccuWeather*, January 18, 2019, https://www.accuweather.com/en/weather-news/can-you-trust-weather-forecasts-during-the-government-shutdown-absolutely/70007173 (accessed March 18, 2019).

31. Jim Eberwine, phone interview with author, January 2019.

32. Dan Sobien, phone interview with author, January 2019.

33. Tom Howell Jr., "Trump's Pick for NOAA Job Withdraws, Citing Health Reasons," *Washington Times*, November 20, 2019, https://www.washingtontimes.com/news/2019/nov/20/trumps-pick-for-noaa-job-withdraws-citing-health-reaso/ (accessed January 23, 2020).

34. Batten, *The Weather Channel*, p. 150.

35. Ken Reeves, interview with author, April 2009.

36. Jason Samenow, "Joe Bastardi Resigns from AccuWeather," *Washington Post*, February 22, 2011, http://voices.washingtonpost.com/capitalweathergang/2011/02/joe_bastardi_resigns_from_accu.html (accessed March 18, 2019).

37. Tom Russell, "Our Guest Meteorologist Joe Bastardi's Winter Prediction," *Channel 21, Harrisburg*, November 1, 2018, https://local21news.com/news/videos/our-guest-meteorologist-joe-bastardis-winter-prediction (accessed March 18, 2020).

Chapter Eight

1. Martin Griffith, Associated Press, "'Father of Snow Surveying' Honored," *Rapid City Journal*, May 4, 2006, p. 14.

2. James E. Church, "The Saga of Mount Rose Observatory," *Scientific Monthly* 44 (February 1937): 141, https://archive.org/stream/in.ernet.dli.2015.534673/2015.534673.The-Scientific_djvu.txt (accessed March 18, 2019).

3. Jim Angel, "Climate of Chicago: Descriptions and Normals," *State Climatologist Office for Illinois*, https://www.isws.illinois.edu/statecli/general/chicago-climate-narrative.htm (accessed January 25, 2020).

4. Spencer Miller, "The History of Church: How NRCS' Snow Survey Program Got Started," *Natural Resources Conservation Service Vermont, U.S. Department of Agriculture*, https://www.nrcs.usda.gov/wps/portal/nrcs/detail/vt/home/?cid=NRCSEPRD329084 (accessed January 25, 2020).

5. "2019 Local Climatological Data Annual Summary with Comparative Data, New York, NY," *National Centers for Environmental Information*, https://www.ncei

.noaa.gov/pub/orders/IPS/IPS-A9AB14AC-E64F-43A3-BF1F-A1B99D648F43.pdf (accessed March 18, 2019).

6. "2019 Local Climatological Data Annual Summary with Comparative Data, Las Vegas, NV," *National Centers for Environmental Information*, https://www.ncei .noaa.gov/pub/orders/IPS/IPS-8BBB3BDA-55FA-4135-B83C-B14C2B1BE38A.pdf (accessed March 18, 2019).

7. "2019 Local Climatological Data Annual Summary with Comparative Data, Denver, CO," *National Centers for Environmental Information*, https://www.ncei.noaa .gov/pub/orders/IPS/IPS-7F08613D-D221-4B1C-98CA-CBAAC62E5084.pdf (accessed March 18, 2019).

8. Ron Aramovich, "How Snow Survey Products and Data Are Used," *National Resources Conservation Service*, December 25, 2006, p. 1, https://www.wcc.nrcs.usda .gov/ftpref/downloads/centennial/article3920061225.pdf (accessed March 18, 2019).

9. "Lake Tahoe: About the Area," *U.S. Forest Service, U.S. Department of Agriculture*, https://www.fs.usda.gov/main/ltbmu/about-forest/about-area (accessed March 18, 2019).

10. Bette Road Anderson, *Weather in the West* (Palo Alto, CA: American West Publishing, 1975).

11. Marc McLaughlin, "The Sierra's Snowiest Winter," *Tahoe Weekly*, November 30, 2016, https://thetahoeweekly.com/2016/11/sierras-snowiest-winter/ (accessed March 18, 2019).

12. "Los Angeles Water Issue: Why It's Not Just the Drought," *University of Southern California Viterbi School of Engineering*, https://viterbi.usc.edu/water/ (accessed January 25, 2020).

13. Aramovich, "How Snow Survey Products and Data Are Used."

14. Chet Henricksen, phone interview with author, January 1996.

15. Chet Henricksen, phone interview with author, January 1994.

16. Anthony R. Wood, "How Deep Was That 1996 Snow, Really?" *Philadelphia Inquirer*, January. 8, 2001, p. C3.

17. James E. Church, *Sierra Club Bulletin*, vol. 6 (San Francisco, CA: Sierra Club, 1906), p. 181.

18. Bernard Mergen, "Seeking Snow: James E. Church and the Beginnings of Snow Science," *Nevada Historical Society Quarterly* 35, no. 2 (Summer 1992): 75–104, http://epubs.nsla.nv.gov/statepubs/epubs/210777-1992-2Summer.pdf (accessed January 25, 2020).

19. Mergen, "Seeking Snow."

20. James E. Church, Snow Surveyors' Forum, Western Snow Conference, 1952.

21. James E. Church, "The Mount Rose Weather Observatory," *Monthly Weather Review* 34, no. 6 (June 1906): 255–61.

22. Jim Marron, "History of Data Collection on Mount Rose, Nevada," *National Resources Conservation Service*, April 3, 2006, https://www.wcc.nrcs.usda.gov/ftpref/ downloads/centennial/article1420060403.pdf (accessed March 18, 2019).

23. Bernard Mergen, *Snow in America* (Washington, DC: Smithsonian Institution Press, 1997).

24. Mergen, *Snow in America*, p. 132.

25. James E. Church, "First Instrument Shelter on Mount Rose," *Scientific Monthly* 44 (February 1937): 141, https://archive.org/stream/in.ernet .dli.2015.534673/2015.534673.The-Scientific_djvu.txt (accessed March 18, 2019).

26. Mergen, *Snow in America*.

27. Nolan Doesken, phone interview with author, January 1996.

28. Miller "The History of Church."

29. Church, "First Instrument Shelter on Mount Rose," p. 149.

30. "The International Commission of Snow," *Nature* 139, no. 63 (January 1937), https://doi.org/10.1038/139063b0.

31. Griffith, "'Father of Snow Surveying' Honored," p. 14.

32. David Laskin, *The Children's Blizzard* (New York, London, Toronto, and Sydney: Harper Perennial, 2004), p. 6.

33. "The Historic Blizzard of March 2–5, 1966," *National Weather Service*, https://www.weather.gov/fgf/blizzardof66 (accessed March 18, 2019).

34. "The Historic Blizzard of March 2–5, 1966."

35. Laura Clark, "The 1887 Blizzard That Changed the American Frontier Forever," *Smithsonian.com*, January 9, 2015, https://www.smithsonianmag.com/smart-news/1887-blizzard-changed-american-frontier-forever-1-180953852/#Mc3rXcA1646LYrDA.99 (accessed March 18, 2019).

36. Laskin, *The Children's Blizzard*, pp. 2–3.

37. "The 'Children's Blizzard' in the Black Hills Country," *National Weather Service*, https://www.weather.gov/unr/1888-01-12 (accessed March 19, 2019).

38. Donald R. Hickey, Susan A. Wunder, and John R. Wunder, *Nebraska Moments* (Lincoln: University of Nebraska Press, 2007), p. 132.

39. Hickey, Wunder, and Wunder, *Nebraska Moments*, p. 139.

40. Hickey, Wunder, and Wunder, *Nebraska Moments*, p. 130.

41. Alexandra S. Levine, "Overlooked No More: Minnie Freeman Penney, Nebraska's 'Fearless Maid,'" *New York Times*, October 3, 2018, https://www.nytimes.com/2018/10/03/obituaries/minnie-mae-freeman-penney-overlooked.html (accessed March 18, 2018).

42. Laskin, *The Children's Blizzard*, p. 147.

43. Levine, "Overlooked No More."

Chapter Nine

1. Macy E. Howarth, Christopher D. Thorncroft, and Lance F. Bosart, "Changes in Extreme Precipitation in the Northeast United States: 1979–2014," *Journal of Hyrdrometeorology* 2, no. 4 (March 2019): 673–89, https://journals.ametsoc.org/doi/abs/10.1175/JHM-D-18-0155.1 (accessed March 19, 2019).

2. Nolan Doesken and Arthur Judson, *The Snow Booklet* (Fort Collins: Colorado State University Department of Atmospheric Sciences, 1997).

3. "Global Climate Report," *National Centers for Environmental Information*, December 2018, https://www.ncdc.noaa.gov/sotc/global/201812 (accessed March 18, 2019).

4. David Robinson, interview with author, January 24, 2020.

5. Bob Taylor, phone interview with author, January 7, 2020.

6. Tyler Malone, "The Road Taken by Robert Frost through New England," *Los Angeles Times*, June 29, 2018, https://www.latimes.com/books/la-ca-jc-robert-frost-20180629-story.html (accessed March 18, 2019).

7. Jeff Leich, "Chronology of Snowmaking, Notes for 2001 Exhibit, New England Ski Museum," *New England Ski Museum*, http://newenglandskimuseum.org/wp-content/uploads/2012/06/snowmaking_timeline.pdf (accessed March 18, 2019).

8. Leich, "Chronology of Snowmaking."

9. Allen Adler, "The Battle of Fifth Avenue," *Skiing Heritage* (March 2003): 17–18.

10. Nils Ericksen, "A Short History of Snowmaking," *Ski Area Management* 19, no. 3 (May 1980): 70–71.

11. Huston Horn, "Vermont's Phenomenal Snow Man," *Sports Illustrated*, March 20, 1961, https://www.si.com/vault/issue/41792/65 (accessed January 28, 2020).

12. Ericksen, "A Short History of Snowmaking."

13. Ericksen, "A Short History of Snowmaking."

14. John Hitchcock, "John Hitchcock Records the First-Ever Man-Made Snow," *Skiing Heritage* (September 1999): 10.

15. "Method for Making and Distributing Snow," U.S. patent no. 2,676,471, filed by W. M. Pierce Jr., December 14, 1950, https://patents.google.com/patent/US2676471A/en (accessed March 19, 2019).

16. Leich, "Chronology of Snowmaking."

17. "Southeast Frosts and Freezes," *North Carolina Climate Office*, https://climate.ncsu.edu/edu/FrostFreeze (accessed March 18, 2019).

18. Leich, "Chronology of Snowmaking."

19. Bob Laylo, "No Snow? That Didn't Stop Jim Thorpe Man," *Allentown Morning Call*, January 19, 1997, p. B5.

20. "John Guresh," *Pennsylvania Snow Sports Hall of Fame*, http://www.pasnowsportsmuseum.com/hall-of-fame/2003/JohnGuresh.pdf (accessed March 19, 2019).

21. Michael Strauss, "Snowmaking's First Fall to Be Hailed," *New York Times*, December 28, 1975, p. 142.

22. Laylo, "No Snow?" p. 5.

23. Strauss, "Snowmaking's First Fall to Be Hailed," p. 142.

24. Strauss, "Snowmaking's First Fall to Be Hailed," p. 142.

25. Strauss, "Snowmaking's First Fall to Be Hailed."

26. "Walter Schoenknecht: Hall of Fame Class of 1979," I. William Berry, nomination letter for Walter Schoenknecht to the U.S. National Ski Hall of Fame,

1979, *U.S. Ski–Snowboard Hall of Fame*, https://skihall.com/hall-of-famers/walter-schoenknecht/ (accessed July 15, 2020).

27. Leich, "Chronology of Snowmaking."

28. Hitchcock, "John Hitchcock Records the First-Ever Man-Made Snow."

29. "Cold and Warm Episodes by Season," *Climate Prediction Center*, https://origin.cpc.ncep.noaa.gov/products/analysis_monitoring/ensostuff/ONI_v5.php (accessed March 18, 2019).

30. "Claire Bousquet, Eastern Area Pioneer," *Skiing Heritage* (December 2003): 30–31.

31. Laylo, "No Snow?"

32. "Big Boulder Ski Area History," *PAskiresorts.com*, http://paskiresorts.com/big-boulder-ski-area-history/ (accessed March 19, 2019).

33. "John Guresh."

34. Ericksen, "A Short History of Snowmaking."

35. "Method for Making and Distributing Snow," U.S. patent no. 2,968,164, filed by A. W. Hansen, January 17, 1961, https://patents.google.com/patent/US2968164 (accessed March 19, 2019).

36. Strauss, "Snowmaking's First Fall to Be Hailed," p. 142.

37. Bob Taylor, e-mail to author, January 20, 2020.

38. Karen D. Lorentz, *Killington: A Story of Mountains and Men* (Shrewsbury, VT: Mountain Publishing, 1990), p. 98.

39. Lorentz, *Killington*.

40. Lorentz, *Killington*, p. 121.

41. Bernard Mergen, *Snow in America* (Washington, DC: Smithsonian Institution Press, 1997), pp. 109–10.

42. "Walter Schoenknecht: Hall of Fame Class of 1979."

43. Huston Horn, "Vermont's Phenomenal Snow Man," *Sports Illustrated*, March 20, 1961, https://vault.si.com/vault/1961/03/20/vermonts-phenomenal-snow-man (accessed March 19, 2019).

44. "Mount Snow," *NewEnglandSkiHistory.com*, https://www.newenglandskihistory.com/Vermont/mtsnow.php (accessed March 19, 2019).

45. Horn, "Vermont's Phenomenal Snow Man."

46. "Mount Snow."

47. Matt Weiser, "How Ski Towns Turn Wastewater into Snowy Slopes," *Pacific Standard*, February 15, 2018, https://psmag.com/environment/turning-wastewater-into-snow (accessed March 19, 2019).

48. "Hopi Tribe v. Ariz. Snowbowl Resort, Ltd.," *Casemine*, November 29, 2018, https://www.casemine.com/judgement/us/5c04adb3342cca1b36c266a9 (accessed March 19, 2019).

49. Mergen, *Snow in America*.

50. Bob Taylor, interview with author, January 9, 2020.

51. Taylor, interview with author.

52. Heidi Davis, "Snow Job: How Hollywood Fakes Winter on Film," *Popular Mechanics*, February 9, 2013, https://www.popularmechanics.com/culture/movies/g1092/snow-job-how-hollywood-fakes-winter-on-film/ (accessed March 19, 2019).

53. Ben Cosgrove, "'It's a Wonderful Life,'" *Life.com*, November 26, 2013, http://time.com/3613817/its-a-wonderful-life-rare-photos-from-the-set-of-a-holiday-classic/ (accessed March 19, 2019).

54. Tristram Coffin, interview with author, December 1983.

55. Stephen H. Schneider and Randi Londer, *The Coevolution of Climate and Life* (San Francisco, CA: Sierra Club Books, 1984).

Chapter Ten

1. "Are You Dreaming of a White Christmas?" *National Centers for Environmental Information, National Oceanic and Atmospheric Administration*, https://www.ncdc.noaa.gov/news/historical-probability-of-white-christmas (accessed March 20, 2019).

2. Roddy Scheer, "Is Skiing Dead Due to Global Warming?" *E-magazine*, November 24, 2018, https://emagazine.com/skiing-dead/ (accessed February 13, 2020).

3. Elizabeth Borakowski and Matthew Magnusson, "Climate Impacts on the Winter Tourism Economy in the United States," *Natural Resources Defense Council, Protectourwinters.org*, December, 2012, https://www.nrdc.org/sites/default/files/climate-impacts-winter-tourism-report.pdf (accessed February 13, 2020).

4. David Robinson, interview with author, January 28, 2020.

5. Robinson, interview with author, January 28, 2020.

6. "Global Climate Report, Annual 2018," *National Centers for Environmental Information, National Oceanic and Atmospheric Administration*, https://www.ncdc.noaa.gov/sotc/global/201813 (accessed March 20, 2019).

7. Deke Arndt, interview with author, September 28, 2016.

8. "Weather Forecasting, Banned by Taliban, Makes a Comeback in Afghanistan," *Canadian Press*, February 19, 2004.

9. Arndt, interview with author.

10. "Arctic Sea Ice Minimum," *NASA*, https://climate.nasa.gov/vital-signs/arctic-sea-ice/ (accessed February 5, 2020).

11. "Southern Hemisphere Snow Cover Extent," *National Centers for Environmental Information, National Oceanic and Atmospheric Administration*, https://www.ncdc.noaa.gov/monitoring-references/dyk/sh-snowcover (accessed March 20, 2019).

12. Herbert Riehl, "Jet Streams of the Atmosphere," *Colorado State University, Department of Atmospheric Sciences*, Report No. 32, https://mountainscholar.org/bitstream/handle/10217/24425/0032_Bluebook.pdf?sequence=1&isAllowed=y (accessed March 20, 2019).

13. Wallace Broecker, phone interview with author, October 1987.

14. Wallace Broecker, "What If the Conveyor Were to Shut Down? Reflections on a Possible Outcome of the Great Global Experiment," *GSA Today* 9, no. 1 (January 1999): 1–7, https://faculty.washington.edu/wcalvin/teaching/Broecker99.html (accessed March 20, 2019).

15. Wallace Broecker, interview with author, August 2005.

16. Broecker, interview with author, August 2005.

17. Robert Gagosian, "Abrupt Climate Change: Should We Be Worried?" *Woods Hole Oceanographic Institution*, February 10, 2003, http://www.whoi.edu/page .do?pid=83339&tid=7342&cid=9986 (accessed March 20, 2019).

18. Peter Schwartz and Doug Randall, "An Abrupt Climate Change Scenario and Its Implications for United States National Security," *Defense Technical Information Center*, October 2003, https://apps.dtic.mil/dtic/tr/fulltext/u2/a469325.pdf (accessed March 20, 2019).

19. D. A. Smeed et al., "The North Atlantic Ocean Is in a State of Reduced Over-turning," *Geophysical Research Letters* 45, no. 3 (January 2018): 1,527–33, https://doi .org/10.1002/2017GL076350.

20. Gagosian, "Abrupt Climate Change."

21. Marilla Steuter-Martin and Loreen Pindera, "Looking Back on the 1998 Ice Storm 20 Years Later," CBC, January 4, 2018, https://www.cbc.ca/news/canada/ montreal/ice-storm-1998-1.4469977 (accessed March 20, 2019); "10th Anniversary of the Devastating Ice Storm in the Northeast," *National Centers for Environmental Information*, January 5, 2008, https://www.weather.gov/media/btv/events/Ice Storm1998.pdf (accessed March 20, 2019).

22. Summer Praetorius, Maria Rugentstein, Geta Persad, and Ken Caldeira, "Global and Arctic Climate Sensitivity Enhanced by Changes in North Pacific Heat Flux," *Nature Communications* 9, no. 3,124 (2018), https://www.nature.com/articles/ s41467-018-05337-8 (accessed March 18, 2019).

23. D. E. Christiansen, S. L. Markstrom, and L. E. Hay, "Impacts of Climate Change on the Growing Season in the United States," *Earth Interact* 15 (2011): 1–17, https://doi.org/10.1175/2011EI376.1.

24. Gilbert M. Gaul and Anthony R. Wood, "Crisis on the Coast," *Philadelphia Inquirer* series, March 6–13, 2000.

25. William M. Gray, interview with author, March 1999.

26. Michael E. Mann and Kerry A. Emmanuel, "Atlantic Hurricane Trends Linked to Climate Change," *Eos* 87, no. 24 (June 2006): 233–44, http://www.cli mateaudit.info/pdf/others/mann.emanuel.eos.2006.pdf (accessed March 20, 2019).

27. Frank Marks, phone interview with author, October 2012.

28. "Sea Level Rise: What Is Expected in New York," *New York City Department of Environmental Protection*, https://www.dec.ny.gov/energy/45202.html (accessed February 6, 2020).

29. Harold Brooks, interview with author, June 2011.

30. Louis W. Uccellini and Paul J. Kocin, *Northeast Storms*, vol. 2 (Boston: American Meteorological Society, 2004).

31. David Robinson, interview with author, January 24, 2020, February 4, 2020.

32. "Transcript of the Interview with Dr. Louis Uccellini, Director of the National Weather Service, 27 February 2013," *Space, Science, and Engineering Center*,

University of Wisconsin–Madison, February 27, 2013, https://www.ssec.wisc.edu/news/transcript-of-the-interview-with-dr-louis-uccellini (accessed March 20, 2019).

33. Judah L. Cohen, Jason C. Furtado, Mathew A. Barlow, Vladimir A. Alexeev, and Jessica E. Cherry, "Arctic Warming, Increasing Snow Cover, and Widespread Boreal Winter Cooling," *Environmental Research Letters* 7 (January 2012): 1–8, https://iopscience.iop.org/article/10.1088/1748-9326/7/1/014007/pdf (accessed March 20, 2016).

34. Stephen Strader, phone interview with author, January 18, 2019.

35. Caitlyn Kennedy, "Wobbly Polar Vortex Triggers Extreme Cold Air Outbreak," *Climate.gov*, January 8, 2014, https://www.climate.gov/news-features/event-tracker/wobbly-polar-vortex-triggers-extsreme-cold-air-outbreak (accessed January 8, 2014).

36. Louis W. Uccellini, phone interview with author, February 4, 2020.

37. Caitlyn Kennedy and Rebecca Lindsey, "How Is the Polar Vortex Related to the Arctic Oscillation?" *Climate.gov*, January 20, 2014, https://www.climate.gov/news-features/event-tracker/how-polar-vortex-related-arctic-oscillation (accessed February 7, 2020).

38. Uccellini, phone interview with author.

39. "Winter 2019/20 Melt/Implode Thread," *Phillywx.com Forum*, January 28, 2020, http://www.phillywx.com/forum/4-obsbanter/ (accessed February 8, 2020).

40. Uccellini, phone interview with author.

41. David W. J. Thompson, "A Brief Introduction to the Annular Modes and Annular Mode Research," *Annular Modes Website*, http://www.atmos.colostate.edu/~davet/ao/introduction.html (accessed February 11, 2020).

42. Jennifer A. Francis and Stephen J. Vavrus, "Evidence Linking Arctic Amplification to Extreme Weather in Midlatitudes," *Geophysical Research Letters* 39, no. 6 (March 2012): 1,000–1,029, doi 10.1029/2012GL051000.

43. Jennifer A. Francis and Stephen J. Vavrus, "Evidence for a Wavier Jet Stream in Response to Rapid Arctic Warming," *Environmental Research Letters* 10, no. 1 (January 2015), https://iopscience.iop.org/article/10.1088/1748-9326/10/1/014005 (accessed March 20, 2019).

44. Uccellini, phone interview with author.

45. Judah Cohen, phone interview with author, February 1, 2020.

46. Mike Halpert, conversation with author.

47. Halpert, conversation with author.

48. Robinson, interview with author, January 24, 2020, February 4, 2020.

49. Caryll Houselander, *Essential Writings* (Maryknoll, NY: Orbis Books, 2005), pp. 182–83.

Epilogue

1. Cindy Stauffer, "Bread and Milk Fly Off the Shelves Faster Than You Can Say Blizzard," *Lancaster New Era*, March 12, 1993, p. 1.

2. Susan Bettencourt, "Local Population Prepares for Fury," *Brattleboro Reformer*, March 13, 1993, p. 1.

3. Kevin Russell, interview with author, February 5, 2010.

4. Brad Tuttle, "Panic Shopping! How a Blizzard Turns Us into Irrational Hoarders at the Grocery Store," *Money Magazine*, January 26, 2015, https://money.com/blizzard-panic-shopping-groceries/ (accessed February 18, 2020).

5. Jere Downs, personal correspondence with author, February 15, 2020, February 21, 2020.

6. IBIS World Staff Member (2134334542), "IBIS World Industry Report 44511, Supermarkets and Grocery Stores in the U.S.," December 2019, p. 25.

7. Cecilia Fernandez, "Supermarkets and Grocery Stores in the U.S.," *IBISWorld*, December 2019, p. 25.

8. Downs, personal correspondence with author, February 15, 2020.

9. Downs, personal correspondence with author, February 21, 2020.

10. Jere Downs, phone interview with author, February 17, 2020.

11. Jaime L. Kurtz, "Panic-Shopping: The Psychology of the Bread-Milk-Eggs Rush," *Psychology Today*, January 23, 2016, https://www.psychologytoday.com/us/blog/happy-trails/201601/panic-shopping-the-psychology-the-bread-milk-eggs-rush (accessed February 14, 2020).

12. Baroclinic Zone, "Panic Room," *Americanwx.com*, December 24, 2019, p. 1, https://www.americanwx.com/bb/topic/52861-panic-room/?ct=1582646120 (accessed February 25, 2020).

13. "Jan. 7–8th Snow Threat," *Phillywx.com*, January 7, 2020, p. 11, http://www.phillywx.com/topic/1607-jan-7-8th-snow-threat/page/11/ (accessed February 21, 2020).

14. "Jan. 7th–8th Snow Threat," p. 5.

15. Robert Frost, *Two Tramps in Mud Time*, ed. Edward C. Lathem (New York: Henry Holt and Company, 1979), p. 277.

16. Ralph Waldo Emerson, "The Snow-Storm," *Poetry Foundation*, https://www.poetryfoundation.org/poems/45872/the-snow-storm-56d22594aa595 (accessed February 22, 2019).

17. Emerson, "The Snow-Storm."

Index

Note: Page references for figures are italicized.